Newcastle upon Tyne

MAPPING THE CITY

Michael Barke, Brian Robson and Anthony Champion

BIRLINN

First published in Great Britain in 2021 by
Birlinn Ltd
West Newington House
10 Newington Road
Edinburgh EH9 1QS

www.birlinn.co.uk

ISBN: 978 1 78027 726 4

British Library Cataloguing-in-Publication Data
A catalogue record for this book is available on request
from the British Library

Designed and typeset by Mark Blackadder

PAGE ii:
This attractive map of Newcastle circa 1851 reminds
us that much historical cartography was a team effort.
The co-operation of John Tallis (1817–1876) as
publisher, John Rapkin (1835–1865) as cartographer
and engraver, and Henry Winkles (1801–1860), an
architectural illustrator, produced this and many other
beautifully illustrated maps in the mid nineteenth century.
The map shows the newly completed High Level Bridge
(1849) and Central Station (1850), whilst the still intact
chares of the central Quayside area to the east of Sandhill
shows that the map pre-dates the Great Fire of 1854.

Printed and bound by PNB, Latvia

Contents

CONTENTS

Simonfide hill   Newbiggin   Wintrigh   W.   Wetherington ca.
Todhorn   Horfeley
Eliesdon   Greneleigh.   Ninni. kirk   Esbley   Vgham   Lynton
R T H   ton   Whitton ca.   Tritlinton   Lyne flu.
Raylees   y Rey   Farnelaw   Stanton   Heborne   Woodhorn
Woodburnes   Pont flu.   Ninriding   New chap.   Bottle ca.
Whelpinton kirk   The Grange   Seaton
des mou th   W: Whelpinton   Wallington   horn   Meldon
owse   Middletons   Hart   Morpeth   Slekbornes
kerinton   Bauuton m.   Wanfpel flu.   Mitforthe
Bolam   Bedlington   Bliches Newke
Swinborn par.   Kirkharle   Whanton   Stannington   Cowpon
chofe ca.   Swinborn ca.   Captheaton   Bellofyfe
Haledon   Kirkheaton   Blich. flu.   Seton dalauel
Cholerton   Ingay   Belfey ca.   Barwik Shotton   Horton
Haugh ton   St Ofwald.   Bingfeld   Madfen   Stauerton   Ogle ca.   Brinck ley   Wetefla des   Sighill
newbrug   Whitten ton   nes   Ponteland   Hartley
B R I A   Diffington Pont fl.   Hallowell
St Iohn Lee   Hawkwell   Darefhall   N. Gosford   Erifden
Warden   Marlow   Preftik   Tynmouth ca.
Hexham   Aydon ca.   Heddon   Picticus murus   Bruntons   Bentham ma
Corbridge   Ouingham   S. Gosford   Wauson
ok   Newbigin   Biwell ca.   Newbrug   Gates head
ridge   Dilston   Prud. do ca.   Riton   Benwel   Tyne flu.   Ierro   Harton
Doteland   Hindley   Elt ringam   Willington   New caftel   Nether Heworth
Slealey   Blackhall Spen   Whikha   Gatefend   Bowdon
Fauterlei   Whittenftall   Hollinfide   Whitborn
Stelchall   Afpersheales   Shotley   Ebchefter   Rauenfwarth ca.   Wafhinton   Hilton ca.
Blanchlande   Medumley   Tanfeld   Lamfley   Munkewermouth
newdon   Darwen flu   Benfeldfide   Vrpethe   Bedik   Weare flu.
Hunsterworth   Edmondbyer   Mugglefwik   Ifeton   Shildraw   Whitwell   Harroton   Wearemouth
Yate   Rowley   Lanchefter   Lumley ca.   Silkefworth   Riop
Knichley   Chefter   Lumley   New bottel   Seham
D V N E L M E N S I S   Langley   Holmfide   Houghton
eren   Cowpigh bell   Afhe   Witton   Morehowff   Duwton
ere flu.   dale   Satlei chap.   Relley   Newton   Eppleton
Stanhope   Frofterley   Wulley   Brandon   Durefme   Piddington
Wulfingham   Shirkley   Wheatley
N I   C V M

Introduction

This is a book about maps and the stories they tell. The nature and purpose of maps has changed over the centuries and maps are creatures of their times. Consequently, in analysing the ways in which Newcastle upon Tyne and Tyneside have been represented cartographically from the past to the present, there is no single interpretive model to apply in all cases. We are concerned with the changing relationship between maps of the city and the area in which it sits and the environmental, socio-economic and political contexts within which such maps were produced. We therefore employ different perspectives in the interpretation and exploration of these relationships in map form.

It is in this spirit that we are unashamedly eclectic in our choice of maps and in our approaches to interpreting them. This book is not a catalogue of the maps of Newcastle and Tyneside, a description of one map after another, rather it is an attempt to put the maps into context, recognising that those contexts change over time. There are many different aspects to 'Mapping Newcastle upon Tyne' and its associated region. The common factor throughout, however, in one form or another is the maps themselves. Maps suggest many different directions to take. We choose to explore a variety of segments of their compass.

The history of any city can only be fully understood in relation to its surrounding area. Although true of all cities, Newcastle upon Tyne demonstrates this relationship more clearly than most. The 'upon Tyne' suffix signals not only a differentiation from other 'Newcastles' but the city's symbiotic relationship with the river upon which it stands. This bond is as self-evident in the cartographic history of the city as it is in its social, economic and political history. Developments in the latter spheres have presented map-makers with numerous opportunities to illustrate its changing fortunes and, consequently, Newcastle upon Tyne and Tyneside possess a compelling and distinctive cartographic history. A unique combination of geology, warfare, politics, trade, invention, entrepreneurship, and local and regional rivalries form the essential backdrop to the maps presented in this book.

This cartographic legacy is a venerable one, and includes a plan of Solomon's Temple of the Tabernacle at Jerusalem (1 Kings, chapters 6–7), produced in the Jarrow Monastery Scriptorium early in the seventh century. Such antiquity is rare, but we examine the work of map-makers from many different eras and backgrounds. These maps provide not only a record of patterns and distributions of features at different time periods,

OPPOSITE. Gerard Mercator, detail from *Northumbria, Cumberlandia et Dunelmensis* (1595) [AUTH] Along with other historical authorities, Mercator renders Gateshead as 'Gatesend' (sometimes also 'Gate-side') whilst transporting 'Gateshead' to the mouth of the Tyne!

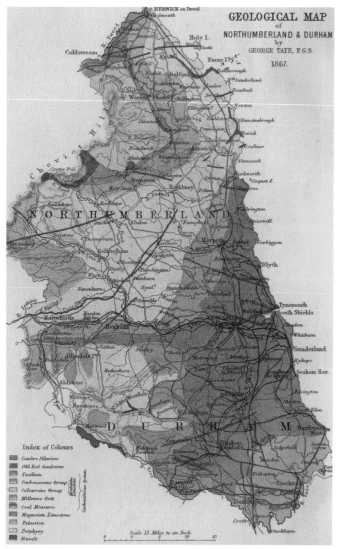

George Tate, *Geological map of Northumberland and Durham* (1867) [SANT] This map appeared in *A New Flora of Northumberland and Durham. With sketches of its Climate and Physical Geography*, vol. II (1867) of Natural History Transactions of Northumberland and Durham, published in 1868.

but many were also aspirational and signalled intent. Although these intentions were not always realised it is fascinating to explore the reasons why and this forms an instructive part of the cartographic record of the area over the centuries. This can be summarised as three broad phases: the early period when

Northumberland could be defined as a 'County of Conflict' and when defence against Vikings, border reivers and Scottish armies was paramount, and when its staunch attempts at 'holding the fort' in the Civil War earned Newcastle the formal motto of *Fortiter defendit triumphans*, bestowed on the town by a grateful Charles I; the exploitation of 'black gold' with the spectacular growth of coal exports and coal mining that literally fuelled a prosperity based on the development of heavy industry; and the economic decline that resulted from the national economy moving from an economic base of heavy industry to one of consumer goods and services. Tyneside, with its heavy reliance on coal mining, shipbuilding, iron and steel, armaments and chemicals, was inevitably a principal victim. Antony Gormley's dramatic Angel of the North that greets travellers to Tyneside is suitably ambiguous, being built of rusting iron, yet its wide-spread arms also signify the promise of regeneration and reinvention within the region.

These broad historical phases have been played out on a physiographical surface of distinctive character. The basic features are clear; location, geology and topography were critical in influencing the nature and form of Newcastle's development. Its location – in the far north of England, yet at the far south of Northumberland – created a strategic function, summarised by William Camden as the 'Eye of the North', and is equally important today. Although the town itself is not emphasised, Mercator's map of 1595 shows Newcastle's regional context clearly. Northumberland was a lawless and dangerous frontier between England and Scotland, the home of the border reivers who plundered on both sides of the border. Successive English monarchs, looking to protect their kingdom from Scottish incursions and fostering their own attempts to invade the Scottish Lowlands, needed a strong defensive base from which to launch attacks. Newcastle was always likely to fulfil that role. Berwick is perched precariously on the border, vulnerable to attacks. Indeed, it changed hands between England and Scotland no less than 14 times in the period up to 1482, since when it has remained English. In contrast, in its early history Newcastle's location was an asset. It became the royal city with its role as stronghold and garrison for border wars and was treated generously by successive monarchs. Today, Newcastle's location is just as critical, although it now works more to the town's disadvantage, especially after the administrative devolution that created a separate Scottish parliament. Put bluntly, Newcastle is the largest English town lying furthest from the interests of London in what is a hugely centralised national economy. It may not be forgotten by London-based politicians, but it is unlikely to rank highly in their priorities. Mercator's map also makes a significant second point, that the east–west links are as important as the north–south. The administrative geographer C.B. Fawcett observed that 'Newcastle . . . is situated at a very important node of natural routes. It is a bridge-town at the head of sea navigation on the Tyne and at the intersection of the two chief routes of the North Country.' While the Great North Road linked the town to both Edinburgh and London, the later Military Road, following the line of the Roman Wall, linked Newcastle and Carlisle.

Geology is the second and arguably most significant influence. While much of Durham and Northumberland are covered with superficial deposits of glacial drift, the underlying solid geology has determined the development of great swathes of Tyneside and its two counties. Tate's map of 1867 shows the striking geological structure to good effect. George Tate was an Alnwick-based tradesman and a local topographer and antiquarian. Three features stand out from his map. Igneous intrusions are one. A great mass of volcanic porphyry forms the Cheviots in the far north of the county, but more dramatic is the intrusive dolomite of the Whin Sill stretching across both Durham and Northumberland. This granitic intrusion is responsible for the ridges of Holy Island and the Farnes, and the crests on which the castles of Bamburgh and Dunstanburgh stand, but for Tyneside the most significant is the line of the Whin Sill on which the central section of Hadrian's Wall was built. It forms a dramatic scarp and guides the Wall on towards Newcastle and its eventual terminus at Wallsend. The second and most significant element of the solid geology is the great expanse of the Carboniferous Coal Measures. The early coal-based pre-eminence of the area was due to outcrops of coal

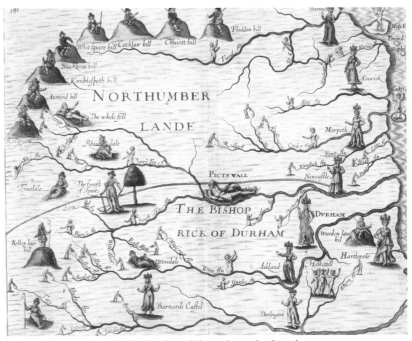

Michael Drayton, *Northumberlande and the Bishoprick of Durham* (1622) [AUTH] This distinctive map was engraved by William Hole who sought to capture Drayton's 'anthropomorphising' and mythical representations of the landscape.

close to the surface and, most critically, within easy distance of the Tyne at a time when overland carriage was cripplingly expensive and access to water transport therefore essential. This export trade was to grow massively in the nineteenth century. The third distinctive geological feature is the wide belt of magnesian limestone that runs diagonally across Durham intersecting the coast in a long stretch between Hartlepool and South Shields. It creates some striking coastal cliff scenery such as the cliffs around Marsden in South Shields, full of caves and isolated stacks and providing nesting sites for hundreds of sea birds, most notably kittiwakes. But this belt profoundly influenced the chronology of exploitation and settlement. The coal measures dip towards the east, underlying the magnesian limestone. So, while the early mines lay in the west and along the banks of the Tyne Gorge, later mining technology enabled deeper pits, spreading increasingly into east Durham and even far out under the sea.

Topography is the third element influencing the area's development. The broad regional context formed by the rivers and topographic features of Northumberland and Durham are charmingly shown on Michael Drayton's 'allegorical' map from the 1622 edition of his *Poly Olbion*. Rivers are represented by nymphs, with small figures pouring water into the headwaters, hills are shown by shepherds, forests by huntresses and settlements by crowned female figures. The 'Picts Wall' is held by a reclining figure. While some details are dubious (the Cheviots are incorrectly shown lying to the north of the Tweed) the map gives an impressive, if romantic, view of the region. A major influence on the region's topography was the Ice Age which spread glaciers south into both counties. One tongue of ice spread into Durham forcing the Wear to turn east to carve out a passage to the sea at what is now Sunderland. Its earlier valley had joined the Tyne at Dunston which was left with the much smaller residue of the Team River which was to provide the site of the area's first industrial estate. Although glacial deposits cover the solid geology in many places, a sandstone ridge constricts the Tyne into a steep-sided channel – the Tyne Gorge – where the modern centres of Gateshead and Newcastle face each other.

This constriction enabled the Romans to bridge the river around AD 122 and build a fort, calling it Pons Aelius. The depth of the gorge was an important feature in the early growth of coal exports since it allowed numerous waggonways to be built that relied in part on gravity to run waggons down to staithes for the export of coal downstream. The action of glacial meltwater cut into the plateau above the Tyne Gorge creating a series of deep waterways, the most significant of which in the development of Newcastle were the Ouse, Pandon and Lort Burns.

As the town spread across the plateau smaller channels were culverted and the valleys filled with rubble to create level surfaces on which roads and buildings could be constructed. The lower part of Grey Street and Dean Street were built on a surface created by cartloads of material tipped into the Lort Burn. The one remaining deep valley is the attractive Jesmond Dene that runs along part of the Ouse Burn from Jesmond south to Heaton Park. Despite these several centuries of landscaping of the central area, its topography can, in the words of Heaton-born Jack Common, socialist writer and friend of George Orwell, be characterised as 'all hills, vales, bridges and one view succeeds another every hundred paces', and in the opinion of David Bean makes Newcastle 'the most visually exciting city in England'.

Tyneside's fluctuating fortunes have been mapped by many cartographers, including some notable local map-makers. Newcastle's earliest accurately surveyed plan was done by James Corbridge in the early eighteenth century. Andrew Armstrong and his son Mostyn fashioned their nine-sheet map of Northumberland in 1769 and were the first to use the title 'Roman' rather than 'Picts' in showing the course of Hadrian's Wall, as well as the first to show that it extended beyond Newcastle to Wallsend. In 1820, John Fryer produced the first large-scale map of the county that showed accurate lines of longitude, but he also surveyed the large map of Newcastle that was published in 1770 by Charles Hutton, and he published detailed maps of the Tyne and plans of local estates. The most prolific local surveyor was Thomas Oliver whose huge plan of 1830 is generally acknowledged as the finest nineteenth-century map of the town and was only one of a long sequence of Oliver's overall maps of Newcastle including many for the town's Corporation.

These were some of the well-known local cartographers, but there was also a host of local surveyors who produced detailed maps of specific areas or for specific purposes. Many of these remain shadowy figures about whom little is known, but their maps in the archives and libraries of Newcastle and elsewhere bear witness to their work and talents. We have tried to select the most appropriate maps to introduce each of the chapters, whose chronology is determined by the date of their main map. Many of the maps that we have unearthed from the archives and libraries are very rare and many were previously unknown to us. There has not been space to use more than a few of them. We hope that our readers will share our excitement and fascination with those that are presented in this book.

Acknowledgements and map sources

We are extremely grateful to Birlinn for its eagerness to include Newcastle upon Tyne in its series of city cartographies and especially to Hugh Andrew and Andrew Simmons for their enthusiastic support and valuable advice and guidance throughout. The production team at Birlinn have been unfailingly helpful and efficient. With Brian Robson's sad death during the course of preparing this book, we are especially grateful to his family, Glenna Ransom and Peter Conway, in facilitating access to his magnificent map collection which forms the central core of this publication. In preparing the maps and text of the book, we have been further helped by many other people, not least Martin Dodge and Nick Scarle of Manchester University's Geography Department, who facilitated and conducted the scanning of Brian's maps, and Terry Wyke of Manchester Metropolitan University, who compiled the book's index. We have also had invaluable support from those in libraries, archives and museums who have helped us locate and scan many of the rarer and more delicate maps that, as a result, we have been able to include in the book. Fundamental to this process have been Alex Healey and Mick Sharp at Newcastle University's Robinson Library, without whose help this volume simply would not have appeared. We have also been incredibly fortunate in the unstinting support we have received from a remarkable group of custodians of, amongst many other things, the region's cartographic heritage: Jen Bell, manager of the Gateshead Archive at Gateshead Central Library; Chris Hunwick archivist at Collections and Archives, Northumberland Estates Archives; Rachel Gill and Lizzy Baker at Tyne & Wear Museum and Archives; Sarah Mulligan and Andrew Scrogham, Head of the Local History Collections at Newcastle Central Library have been unfailing in their courtesy and patience in the face of our stumbling efforts to explain precisely which maps we required. No less generous with their time and expert advice have been Paul Ternent, Commercial Development Manager at the Northumberland County Archives, Ashington; Jennifer Hillyard, Library and Archives Manager at the Mining Institute; Kay Easson, Librarian at the Literary & Philosophical Society; Chris Fleet, Map Curator at the National Library of Scotland; Mike Greatbatch, formerly Heritage Manager at the Ouseburn Partnership and Newcastle City Council; and Denis Peel, Librarian at the Society of Antiquaries of Newcastle upon Tyne. We are also particularly grateful to a number of individuals who have assisted in sourcing specific maps, especially Joyce Marti at 'Discover', North Tyneside Central Library; Julian Harrop, Head of Collections at Beamish Open Air Museum; Paul Greenhalgh of Northumbria University's Department of Architecture and Built Environment; Charlotte Barke for introducing us to Open Street Maps; Iain Garfield at Newcastle University Estates Office; Liz Lambert of Nexus; Sam Atkinson at Love to Run Art; and the artist John Coatsworth.

Norman McCord, Emeritus Professor at Newcastle University, was kind enough to read through an early version of the manuscript, and useful advice and comments were also received from Alan Townsend of Durham University, John Goddard and Mike Coombes of Newcastle University's Centre for Urban & Regional Development Studies, and Chris Stephens, formerly at Newcastle City Council. Finally, we are – as ever – in debt to our wives and families for their patience and support which has been especially remarkable in the unique circumstances resulting from the Covid-19 pandemic.

Map sources

The source of each map is shown, in square brackets, at the end of the map captions in each chapter. The following abbreviations are used:

AUTH	Authors' collections
BL	British Library
GAC	Government Art Collection
GLH	Gateshead Local History
LIT & PHIL	Literary & Philosophical Society of Newcastle upon Tyne
NA	National Archives
NCA	Northumberland County Archives
NCL	Newcastle Central Library
NCC	Newcastle City Council
NERA	North East England Regeneration Archive
N.EST	Northumberland Estates
NLS	National Library of Scotland
NTCL	North Tyneside Central Library
NUFC	Newcastle United Football Club
OSM	Open Street Map
ROB	Robinson Library, Newcastle University
SANT	Society of Antiquaries of Newcastle upon Tyne
TWA	Tyne and Wear Archives
TWM	Tyne and Wear Museums

Brian Robson: An appreciation

The name Robson is a 'give-away'. Although an unsentimental individual, Brian's deep attachment to his native Northumbria was evident in any prolonged conversation and shines through clearly in this book. Sadly, *Newcastle upon Tyne: Mapping the City* is his final published work, but he would have regarded this as entirely appropriate. Although most of his life was spent outside north-east England, he was inherently proud of his roots within the region and unfailingly meticulous in recharging his knowledge base on its changing trajectories. In an uncharacteristically self-reflective essay, Brian recalled the (human) geographical questions arising from his daily train journey to the Royal Grammar School in Newcastle from his birthplace and residence in South Shields. Furthermore, many of his fondest personal memories (when pressed) related to rural Northumberland, the location of frequent family holidays and expeditions. According to his older sister, Marie, on these outings, 'our father made sure we could read an O.S. map accurately from a young age.' These experiences instilled Brian's deep interest and fascination with maps, not just as a medium for locating places or finding one's way, but also as an expression of the distribution of human activity, of modes of representation, and of social and economic power.

Brian Robson was one of the leading human geographers of his generation. After his first degree and doctoral thesis (on Sunderland) at Cambridge, he held posts at Aberystwyth University and at the University of Chicago, then returned to Cambridge before being appointed professor of urban geography at Manchester University. There Brian played a major role in designing a geographical dimension to the Single Regeneration Budget (introduced in 1994), which – contradictory to the Thatcherite perspective that social and economic disadvantage was limited to 'those Inner Cities' – recognised that the same problems were present in peripheral housing estates and small towns in areas of economic decline. This became the fundamental tool for applying a more geographically aware and integrated approach to area-based regeneration.

In parallel with this prominent and highly respected academic background, Brian also had an equally distinguished 'hobby'. This was a fascination with the history of cartography, evident in his re-appraisal of John Wood, a long-forgotten but highly significant cartographer, and in his masterly evaluation of the machinations and inner workings of the key actors who produced – and mapped – Newcastle's magnificent 'new' centre in the 1830s and 1840s.

In this sense, Brian's career came through a geographical circle, and this book represents a concluding part of that journey. As co-authors, we hope that it does justice to that schoolboy's daily train journey-based musings on place, people and history – some of the many dimensions of which are mapped and discussed in the following pages.

Michael Barke and Anthony Champion

ada

Pelagus vas
tissimum et
innuum

partia palustris 7 inuia pecudibs 7 pastorbs apta

Ka habden

Aberbzopot

Suthernelande

Regio mon
tuosa 7 nemo
rosa gentem occidet
generans 7 pastora
castm lem sub maris
dinkeual cum 7 harena
dinetum

Brachym
ma
ris

Dunbum

Dunde

Ozk
Inr

ciuitas
sca Andree

isula
colum
uil
li

SCOCIA: ULTRAMARINA

hec et albania dicta est

fit

tyrenie

pars ma
ritima 7 gens
montana

estiuelin

Dunsine
lin

pong

ciuitas
Regine

ciuitas
comi
tis

Galewe
ia

fluuius facies Cludesdale. Dos

edene
burg

Regio scotoz continuoz

Glascu

tuedesdat

murus diuidens scotos 7 pictos olim

Berewse

Engleseta
Insula

adm

North
Wallia

Et clud
clude

murus diuidens anglos 7 pictos olim

Ro
melve bef
burg

Rokef
isula

Kohotes

Roch

North
W.

Karleola

chi

Walte

Snaudun.

Ban
gor
ep.

tuee est

comitat cest

El tyne

Tindale

Tunemue

ayarelud

Cest
a

Houu castm

mon

pinu
mon

Sabina fl. qd 7
maro di. tp sui
excellucam.

Richemud

Stei

Werdale

Duo brach

Wer

Blac

Bev
land

amenuia
s. q dauid
epatus

WALLIA

Duncmui

la

Babe

gens hui regois
de galo bruti
7 pagaui
landas
epatus

ogof gouue

rola ad

mon bise
burne

Heram dra
spi incolas su oske fluuu

Sale
rest

Alurona

Eborduu

karmerdin
ciuitas olim

et britannia di

Pons Burg

Be v.

hube

uerlam

pons fe

em
meslu

c.1255

Building a Northern defence

> Newcastle upon Tyne . . . began her long and eventful history as a frontier station, a role she continued to fill for thirteen centuries after the fall of Rome.
>
> S. Middlebrook, *Newcastle upon Tyne: Its Growth and Achievement*, 1950

Matthew Paris's 'map' of Britain (one of four versions that he drew) offers a striking image of the line of Hadrian's Wall. Based in his monastery at St Albans, Paris was an outstanding cartographer who drew on information from travellers and others; for example, his eastward extension of Scotland undoubtedly derived from Ptolemy. While his map bears scant relationship to the shape of Britain – as is dramatically evident in the depiction of the Scottish Highlands being joined to the mainland by a land bridge south of Stirling – it identifies more than 250 towns and offers a decipherable representation of the rivers and coastline of the country. The 73 miles of Hadrian's Wall are shown as a battlemented line, as is the Antonine Wall to the north (even though that wall was merely built of turf and had long since disappeared).

Hadrian's Wall is invariably associated with the wilds of the Northumberland moorlands with its dramatic switchback along the line of the granitic outcrop of the Whin Sill. Newcastle is rarely, if ever, seen as one of the keys to the Wall. Yet Tyneside played a critical role both in determining the alignment of the Wall at its eastern end and in its early exploration and excavation.

The first printed map of Northumberland (*c.*1576) by Christopher Saxton showed the line of the Wall as it approached Newcastle, dipping south towards the Tyne. But

---

OPPOSITE: Matthew Paris, detail from *Map of Great Britain* (1255–59) [BL] Northern England and Scotland from Paris's elaborate second map of Britain. Unlike a number of later maps (e.g. Saxton and Speed) Paris correctly ends the 'Pict's Wall' at Wallsend, although he does not locate this directly on the Tyne.

John Speed, detail from *South Northumberland with Roman Wall* (1611) [AUTH] Speed produced visually attractive maps but his confident placement of ancient sites is unreliable; Vindomora is near Ebchester 12 miles south-west of Newcastle and although the historian Bourne discusses a possible Gabresentu(m) at Gateshead that name does not appear in modern accounts of the Wall except at Moresby in Cumbria.

John Speed's map of the county (1611), published in his *Theatre of the Empire of Great Britain*, depicts the Wall much more clearly. Reflecting the state of knowledge at the time, neither Speed nor Saxton show the eastern extension of the Wall to Wallsend where it entered the large Roman fort Segedunum at its western gate and left via its south-east corner to drop down to the Tyne. Speed was a diligent researcher for his maps, but his mistaken placement of Segedunum several miles to the north of its true location suggests he may have been misled by local sources. He is unlikely to have simply taken a guess. Further east of Wallsend the river was sufficiently wide not to need a defensive wall, and the fort commanded a clear view along the long straight reach of the river to beyond Jarrow.

The section of the Wall from Newcastle to Wallsend was built slightly later than the stretch west of Newcastle and, unlike the ten-foot-wide 'broad' structure in the west, the Newcastle–Wallsend section had a width of seven and a half

feet. The original wall ended at the fort of Pons Aelius which was built on the high promontory on which the medieval castle was later erected. The fort provided protection for the Roman bridge across the Tyne. Had the Wall gone straight to Wallsend, its alignment would have taken it well to the north of the centre of Newcastle. The exact lines of the Wall and of the fort in Newcastle are still uncertain (both inevitably suffered from pilferage of their stones taken for the construction of the growing town, especially in the later eighteenth and nineteenth centuries). Small stretches of the Wall have been unearthed when new developments in the town entailed major earth-works. For example, a recent discovery was a stretch of the Wall lying outside the Mining Institute in Neville Hall on Westgate Road, and the Lit & Phil building has a plaque identifying a small section at its base. Collingwood Bruce's 1847 map gives an indication of the differences between the views of archaeologists in the absence of detailed excavations. The earlier views of John Horsley, in his *Britannia Romana* of 1732, show a substantially different alignment both of the Wall and of the fort than does Collingwood Bruce's *Guide to the Castle of Newcastle upon Tyne*. Beilby's map (*see* 1788) actually suggests (mistakenly) that there were two walls, one attributed to Severus and the other to Hadrian. More recent discoveries indicate that it appears most likely that the Wall entered the town via what became Westgate Road, joined the large garrison fort that overlooked the river and left along the line of Shields Road to head for Segedunum.

In the early eighteenth century there was considerable above-ground evidence that could be called upon. The antiquarian William Stukeley travelled to the north of England to follow the course of the Wall from Carlisle to Wallsend in 1725. He recorded large stretches of visible wall and ditch. He noted that, leaving Newcastle via Pandon Gate, the Wall was 'very plain thither from Sandgate mill, both the ridge of the wall, and ditch . . . it passes a very deep valley at Euxburn [Ouseburn] . . . The foundation of the wall is yet intire [sic] within the pastures, and a considerable ridge of it is left.' He also noted that both east and west of Newcastle numerous coal mines had been sunk on or close to the line of the Wall, one of

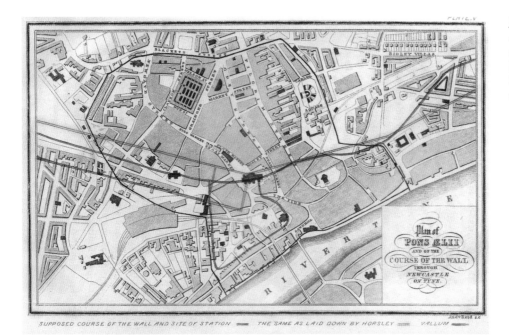

John Collingwood Bruce, *Plan of Pons Aelii and of the course of the wall through Newcastle upon Tyne* (1851) [SANT] Bruce's attempt to outline the boundaries of 'Roman Newcastle' is sensible but conjectural.

which had caught fire and 'vomits out smoke continually, like a volcano'. Leaving Newcastle via the West Gate, he also observed the third major garrison fort on Tyneside, Condercum at Benwell, which 'commands a great prospect every way'.

Newcastle played a significant role in the early exploration and excavation of the Wall through the work of John Clayton, the formidable town clerk who was so influential in facilitating the nineteenth-century remaking of Newcastle. Clayton came from a wealthy family and was town clerk for 45 years from 1822, having succeeded his father Nathaniel who had himself held the post for 37 years. In 1796, Nathaniel acquired a mansion and estate which included the site of the Roman fort of Chesters. John Clayton developed a fascination for the Roman remains in its environs. From the 1830s he bought up land in order to protect and restore parts of the Wall, including Housesteads and Vindolanda, as well as Chesters. He undertook some of the very earliest excavations of the Wall and published numerous articles on his findings in *Archaeologia Aeliana*.

The major Roman site on the south of the Tyne was Arbeia, a large fort built on Lawe Top at South Shields. This commanded the final reach down to the river mouth, though this was probably supplemented by a signal station at the Tynemouth headland, so the riverbanks left unprotected by the Wall were more than adequately 'policed'. It is probable that there was an adjacent port at Arbeia, providing a landing site for troops and for the dispatch of supplies of grain, as this was not only a defensive site with a full garrison but also a general supply base for the Wall. Details of the fort were little known until recently because the site had been covered with houses by the end of the nineteenth century and even in the twentieth century remained a small undistinguished enclave until the housing was cleared in the 1970s. Now, however, it is one of the most comprehensively excavated Roman forts, with a reconstruction of the gatehouse, barracks and Commanding Officer's house. The gatehouse houses a museum displaying artefacts and details of the history of the fort.

Even though the Roman imprint on Tyneside is almost invisible, it was nevertheless structurally important, the fort prefiguring Newcastle's medieval castle and the Roman bridge providing the line and foundations for later crossing points of the Tyne.

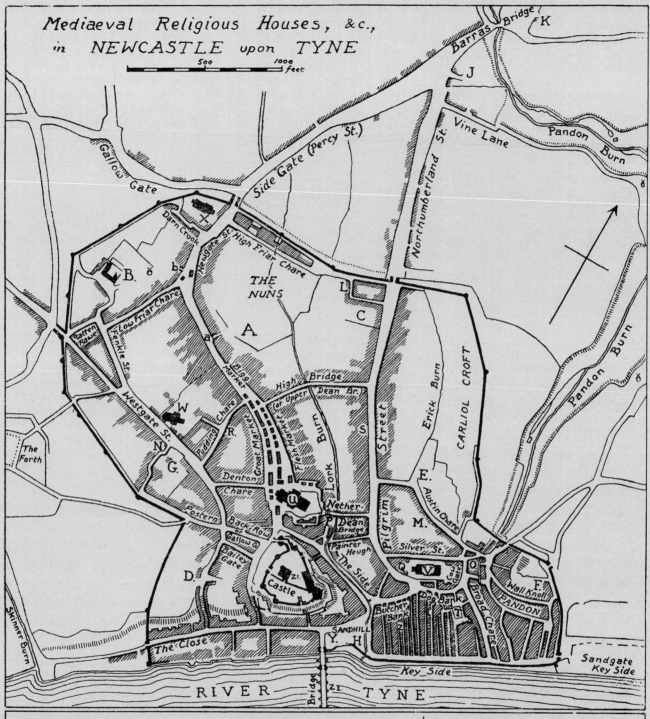

Mediaeval Religious Houses, &c., in NEWCASTLE upon TYNE

500   1000 feet

Gallow Gate
Side Gate (Percy St.)
Barras Bridge
K
J
Vine Lane
Pandon Burn
Northumberland St.
Darn Crook
Newgate St.
High Friar Chare
X
L.
THE NUNS
C
Low Friar Chare
Fenkle St.
Rattenrawe
B.
δ
b.
A
a.
Carliol Croft
Erick Burn
Pandon Burn
Westgate St.
Bigg Market
High Bridge
(or Upper Dean Br.)
W
Pudding Chare
R.
The Forth
N.
G.
Denton Chare
Groat Market
Flesh Market
Lort Burn
S
Pilgrim Street
Austin Chare
E.
Postern
Back Row or Gallow G.
U.
P. Dean Bridge
Nether-
Nether-
M.
Silver St.
Bailey Gate
D.
Castle
z1.
Painter Heugh
The Side
Q.
V
Cow Gate
O.
Wall Knoll
F.
PANDON
Butcher Bank
Dog Bank
Broad Chare
T.
Skinner Burn
The 'Close
SANDHILL
Y. H
Bridge z2
Key Side
Sandgate Key Side

RIVER TYNE

A. St. Bartholomew's Priory.	L. Brigham's Hospital.	X. St. Andrew's Chapelry.
B. Dominican Friars.	M. Ward's Hospital.	Y. St. Thomas M.-on-the-Bridge.
C. Franciscan Friars.	N. 'Spital' Alms House.	z. St. John's Chapel (supposed).
D. { Carmelite Friars after 1307.	O. Stockbridge Alms House	z1. Castle Chapel & Chantry.
{ Friars of the Sack till 1307.	P. Nether Dean Br. do.	z2. Hermit's Chapel-on-the-Bridge (supposed).
E. Austin Friars.	Q. Nykson's do.	
F. { Carmelite Friars till 1307	R. Pudding Chare do.	a. Nuns' Gate.
{ Trinitarians after 1360	S. Pilgrims' Inn.	b. Black Friars Back Gate.
G. St. Mary V. Hospital.	T. Trinity House.	
H. St. Katherine Maison Dieu.	U. St. Nicholas Par. Church.	Compiled from Plans by Speed, in Gardner, Beilby, Thompson, & Longstaff, by R. Neville Hadcock
J. St. Mary Magdalene Hospital.	V. All Saints Chapelry.	1937
K. St. James Chapel & Lazar Ho.	W. St John's Chapelry.	

c.1260

Monkchester

It was a sign of the growing wealth and importance of the town that so many religious orders set up foundations there.

A.W. Purdue, *Newcastle: The Biography*, 2011

Some historians of Newcastle have claimed that in its earliest years the town was called 'Monkchester', reflecting its important roles as a home for monastic establishments and as a place of safety for monks from the wider region in times of invasion by Danes or Scots. Regardless of whether this name was ever used, there can be no doubt about the pivotal role that Tyneside and its wider region played in the early spread of Christianity. Early in the seventh century, the Irish monk St Aidan founded the Lindisfarne monastery on Holy Island, which became a key centre of learning and evangelism, and spread Christianity across Celtic Britain. In 627 Edwin had become the first Christian Northumbrian King, and at Lindisfarne St Cuthbert became the most significant and revered abbot of the monastery.

There were other important Celtic monastic houses in Tyneside and the wider region. The priory at Tynemouth (*see* 1545a) was first established in the seventh century and was the burial place of two of Northumbria's Celtic kings. Of greater moment was the twin monastery founded on the banks of the Wear at Monkwearmouth and of the Tyne at Jarrow. Its particular significance is its link with the Venerable Bede who was initially sent to Wearmouth at the age of seven and then to the newly founded Jarrow monastery in *c.*672. If Lindisfarne is forever associated with the richly illustrated Gospels of the early eighth century, St Paul's at Jarrow is equally famed for Bede's *Historia Ecclesiastica*, providing a broad account of early Celtic and Saxon Britain, and leading to him

---

OPPOSITE: Neville Hadcock, *Plan of Medieval Newcastle upon Tyne showing religious houses* (1937)
[ROB] R.N. Hadcock (1895–1980) was born in Newcastle and became a distinguished historian of medieval ecclesiastical buildings.

becoming known as the father of English history. His many publications encompassed an amazing range of issues: the calculation of the date of Easter, popularising the use of Anno Domini in place of regnal years in chronology, the first Anglo-Saxon record of Hadrian's Wall, scientific work on the cosmos and numerous theological works. His *The Reckoning of Time* has a particular significance for geography since it included work on the relationship between tidal changes and the movement of the moon. For this, Bede is believed to have measured tides in the River Don, the small stream that flowed close to the monastery and into Jarrow Slake to its east.

Saxon monastic buildings tended to be sited close to the sea or be readily accessible by river, reflecting the primacy of sea travel at that time. The unforeseen consequence was that they were prone to frequent attacks by invading Danish marauders. An early raid was in 793 when a party of Vikings attacked Lindisfarne and massacred many of its monks. Tynemouth

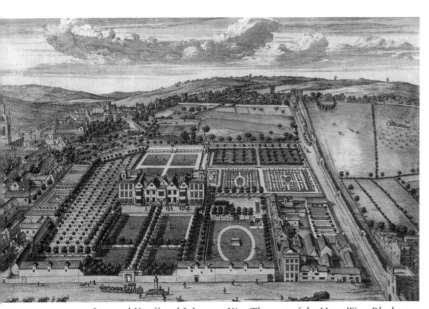

Leonard Knyff and Johannes Kip, *The seat of the Hon. Wm. Blackett with part of the town of Newcastle upon Tyne* (1707) [GAC] Johannes Kip specialised in bird's-eye drawings of English country seats, demonstrated here in the urban setting of Newcastle's northern fringe with Anderson Place and the grounds of the former nunnery of St Bartholomew's to the west.

Priory was plundered by Vikings in 800 and, even though the fortifications were sufficiently strengthened to hold the site against a further attack in 832, the church and monastery were destroyed by the Vikings in 865, and nuns who had sought shelter there were slaughtered. The priory was again plundered in 870. In 876 a Danish chieftain, Halfdan, destroyed the monastic houses at Wearmouth, Jarrow, Lindisfarne and Tynemouth, as well as those in Newcastle. Monks and nuns were massacred, and it was only in the Norman period that the area re-established a significant religious presence.

The Jarrow monastery was reconstructed by a Mercian monk, Aldwin of Winchcombe, after he visited the ruined site in 1074, and he later undertook the renewal of the one at Wearmouth. But it was Newcastle that saw the greatest flowering of religious houses. While there are no contemporary manuscript maps of the town in the Saxon or Norman periods, the outline of religious houses in the mid thirteenth century was valuably reconstructed by Neville Hadcock in 1937. Newcastle was broadly divided into a secular riverside area and a 'religious' area with the friaries and other religious premises on the upper plateau. Newcastle contained an unusually wide range of monastic orders, including Dominicans (Blackfriars) near Stowell Street, Franciscans (Greyfriars) at the top of Pilgrim Street, Augustinians (commonly abbreviated to 'Austins') to the east of Pilgrim Street and Trinitarians in Pandon. In addition, Carmelites (Whitefriars) initially located in Pandon on the site later used by the Trinitarians and then moved to the bottom of Westgate Street where Hanover Square was subsequently built on a site that had earlier been used by the Friars of the Sack (so called because of their simple clothing). As well as the various chapels, the town also had two major hospitals run by monks and nuns, St Mary the Virgin in Westgate Road and St Mary Magdalen at Barras Bridge. All this amounted to a rich spiritual culture in medieval Newcastle and one that was not limited to the town only: most of these religious houses were the recipients of frequent bequests of money and commodities from both leading burgesses and the elite gentry of the wider region.

As the plan of medieval Newcastle shows, most of the

W.H. Capone and John Preston Neale, *Ruins of Jarrow Monastery* (1835) [NCL] Engraved by Capone from an original drawing by the architectural draughtsman, John Preston Neale.

religious houses occupied peripheral sites within the walls and some owned land extramurally. However, in terms of the future development of Newcastle, the most significant of the religious houses was the nunnery of St Bartholomew. Its large site later provided development opportunities for the Georgian reconstruction of the town. The nunnery initially flourished but went through uncertain times in the fourteenth century (when the prioress was accused of 'intrusion, dilapidation, incontinence, and other crimes'), recovering later but then gradually declining. In 1513, it granted the Corporation of the town a hundred-year lease on Nuns Field and by the time of its dissolution in 1540 it merely comprised a prioress and nine nuns. Such waning was typical of all the religious establishments at this time. The growth of industry and commerce, allied with the scandals that often plagued religious houses, led to the major sixteenth-century revolution that saw the old order of dominance by the Church and barons being replaced by entrepreneurs and businessmen. For the ecclesiastics, the major blow came with the dissolution: by 1540 all the major religious houses had been dissolved. By then, though, most had few

inhabitants left – the Dominican friary had the most, although comprising only a prior and twelve brothers, and the Trinitarians had the least, with a solitary warden.

The impact of dissolution on the town was considerable. Most of the land and buildings were sold to merchants or the Corporation. The King kept only the Austin friary, used on rare occasions as the headquarters of the Council of the North when it met outside York. Much later it was to be remembered in the naming of Manors Station. Other friaries were bought by the Corporation and leased to some of the town's guilds – including cordwainers, joiners and smiths – with a particular concentration at Blackfriars. But the major impact was the sale in 1580 of Greyfriars and the nunnery of St Bartholomew's, together with its extensive grounds, to Robert Anderson, a wealthy local merchant. He built an imposing house, which Speed depicted as 'Newe House' on his 1610 plan of Newcastle and which was later called Anderson Place, although it came into the ownership of the powerful Blackett family. It was the nineteenth-century sale of this house and its land that was the key to the town's Georgian reconstruction.

Tinmouth Abbey

Tinmouth Castell

TYNE FLVMEN

1545a

Early Tynemouth: 'a place moste apte and needful to be defended'

Tynemouth priory . . . is one of those intensely evocative northern sites that is both visually dramatic and deeply imbued with the historic early days of Christian missionary activity.

R. Fawcett, 'The Architecture of Tynemouth Priory Church',
in J. Ashbee and J. Luxford, *Newcastle and Northumberland:
Roman and Medieval Architecture and Art*, 2013

The magnesian limestone headland, upon which the ruins of Tynemouth Castle and Priory stand, forms an impressive entrance to the Tyne. Its strategic location overlooking the river mouth, together with its natural defensive capabilities, made it an obvious site for a castle, but the first buildings to occupy the land were religious ones. Sacked by Danish raiders in 800 and 875, it was only in 1083 that they were fully restored and then, due to a quarrel between Robert Mowbray, Earl of Northumbria, and the Bishop of Durham, ownership was transferred to the Benedictine house of St Albans. Although many associated buildings are no longer visible, the current ruins provide a magnificent example of mainly thirteenth-century Gothic architecture. Despite numerous challenges the monastery survived until its dissolution in 1539. Although the ruins of the priory now dominate the castle remains, it is the combination of the two which creates the character of the site. Initial permission to fortify the area with a tower and curtain wall was granted in 1296, with the gatehouse being built in 1390. Religious and military functions were therefore combined.

The 1545 map was probably a product of concerns over a possible French invasion, potentially using south-eastern Scotland as a platform. It is credited to Gian Tommaso Scala, working under the military engineer Sir Richard Lee, and

OPPOSITE: Gian Tommaso Scala, *a coloured plan of Tynemouth Abbey and Castle* (1545) [BL]
Scala's elaborate defences were not executed and a long earth wall and ditch were built instead.

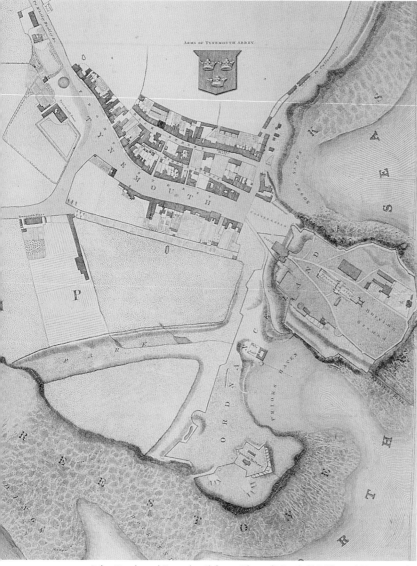

John Rook and Son, detail from *Plan of North Shields and Tynemouth* (1827) [SANT] Detail of Tynemouth village and headland.

pragmatic requirement to fortify the east coast appears to have outweighed any concerns about employing fortification experts from a Catholic country. Their influence is still apparent in Sir Richard Lee's design of Berwick's magnificent Elizabethan walls around 1560. These walls have Italian-style defences with five large bastions.

Tynemouth Castle achieved prominence through its role in Act 2 of Christopher Marlowe's play *Edward II*. The fugitive King took refuge there with his favourite, Piers Gaveston, before fleeing by ship to Scarborough Castle. During the Civil War the castle was captured by the Scots in 1644 and by the Parliamentarians in 1648. Prior's Haven, between the promontory and the Spanish Battery, is clearly marked on Scala's map, and the vessel moored just offshore indicates its role as the place of loading and unloading cargoes to and from the monastery. It later became a fashionable bathing place. The civil settlement of Tynemouth is shown as huddled up against the western walls of the castle but with three roughly parallel streets clearly visible. A tower with warning lights stood on the promontory and was replaced by a lighthouse in 1665 (rebuilt in the eighteenth century and, appropriately enough, coal-fired until 1802). The lighthouse was demolished in 1898 after its replacement by the one on St Mary's Island, north of Whitley Bay.

The priory also played an important commercial role, although this took place within the continuous shadow of competition and legal disputes with the Crown and the burgesses of Newcastle. The thirteenth century was a time of economic growth, as reflected in the magnificence of its new church. The priory had certain rights to produce and obtain commodities, but most of these were concerned with its own subsistence. It seems that successive priors stretched these rules somewhat, exploring several means of commercial expansion. Newcastle burgesses objected strongly, and their petition to Parliament demanded that piers at North Shields should be destroyed. The market was prohibited in 1290 and replaced by an annual fair in 1304, but, following protests from Newcastle, this too was rescinded. Nevertheless, over the next 200 years there was a gradual increase in economic activity at

employed by both Henry VIII and Elizabeth I. The international scale of operation and influence of what must have been a relatively small professional group is indicated by the fact that Lee had taken Scala and another Italian military engineer, Antonio de Bergamo, to Scotland during the war that followed the break with the Catholic Church. Somewhat ironically, the

Tynemouth and its new settlement of North Shields, as religious houses generally became ever more powerful forces. North Shields had been established by Prior Germanus in the early thirteenth century as a base for fishermen, whose hut dwellings were known as 'shiels'. Like most similar religious houses, in the fourteenth and fifteenth centuries Tynemouth Priory became a much more commercial undertaking, for example owning coal mines in Elswick and Benwell, and succeeding in buying out its freehold tenants and letting the holdings at economic rents. The priory made over 100 such purchases and acquired over 2,000 acres, becoming by far the wealthiest religious house in Northumberland with an annual value of £750 in the early 1500s, more than twice as wealthy as Alnwick, Hexham and Newminster whose incomes ranged between £200 and £300 each.

The earliest documentary evidence of the settlement of Tynemouth dates from the late twelfth century and it seems likely that this settlement was then very recent. Archaeological evidence has demonstrated the existence of a cemetery, still in use in 1170 and extending right across the north and south sides of Front Street. It is unlikely that the settlement would have been laid out before the cemetery ceased to be in use. The Lay Subsidy Roll of 1296 names 15 tenants in this secular settlement. Poorer people were exempt from this tax, but some of the recorded occupational names such as 'Richard of the bakehouse' and 'Geoffrey of the brewhouse' suggest an active village. Places of origin in some surnames indicate in-movement from Wylam and Backworth. By 1336 the village contained 117 houses. It developed along the route westwards from the priory and castle, but its morphology (now transformed due to the insertion of Hotspur Street and Percy Park Road running in a north–south direction) suggests that the growth was not haphazard. A degree of formal planning is indicated by the three roughly parallel streets, called Front Street, Middle Street and Back Street (now Percy Street), which are clear on Rook's 1827 map. Morphologically, this is remarkably similar to the original layout of Newcastle's Groat Market, Cloth Market and former Middle Street area (the latter obliterated when Newcastle's Corn Exchange and 'New' Town Hall were built

Robert Goadby, *A view of Tynemouth Castle in Northumberland* (1776) [AUTH] From Goadby's *A New Display of the Beauties of England*.

in the mid nineteenth century). Relatively shallow plots were created fronting onto these three streets, although in Tynemouth Middle Street is truncated at its western end, creating deeper plots extending between Front Street and Back Street, now interrupted by the creation of Percy Park Road running northwards to the sea. The majority of rural settlements in north-east England consisted of a two-row structure, even if these rows were separated by a wide village green. The original structure of Tynemouth was clearly different, and the existence of Middle Street suggests a morphological form intended for a different purpose. Like other 'multiple row' settlements, it seems likely that the intention was for Tynemouth 'village' to develop as an important marketplace. But as late as the 1820s some plots were still underdeveloped, especially in the eastern part of the settlement. The riverside location and the functionally much more flexible site of North Shields allowed the latter to usurp its mother settlement in economic significance.

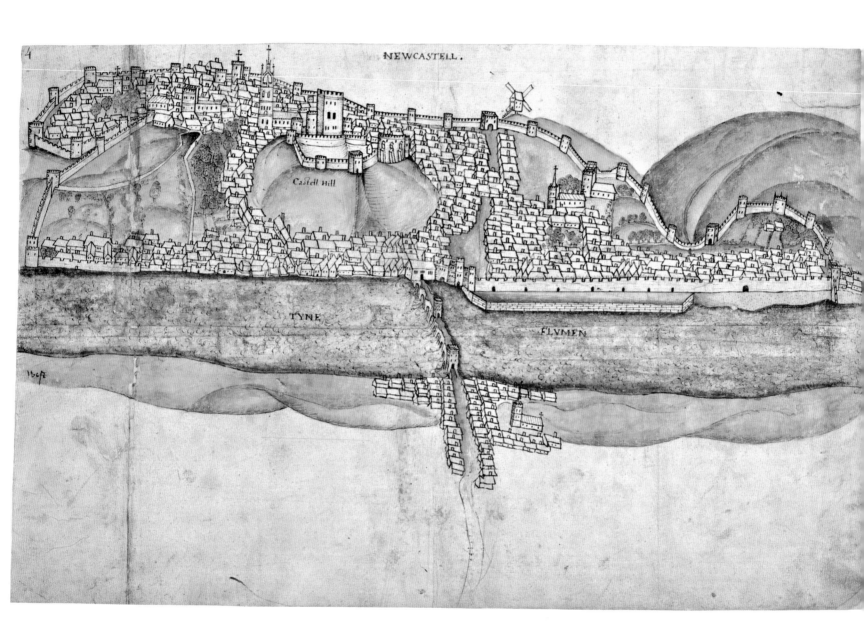

NEWCASTELL.

Castell Hill

TYNE
FLVMEN

West

1545b

Town defences

Border warfare seems to have helped rather than hindered the development of Newcastle.

R. Howell, *Newcastle upon Tyne and the Puritan Revolution*, 1967

Medieval Northumberland was a rough and lawless place. Frequent wars between Scotland and England brought incursions across the border from both sides. This, together with the constant threat of border reivers – mounted gangs of thugs and impoverished locals comprised of both English and Scottish men – meant that protection from violence was necessary for most residents. The array of castles in the county and the plethora of bastle houses (two-storey buildings with livestock on the ground floor and people on the first floor) and pele towers (three-storey fortified houses) show the dangers present from the thirteenth to the early seventeenth centuries.

For English kings, the significant challenges came from the recurring wars with Scotland. Newcastle's strategic importance is signalled by the fact that the first three Edwards visited the town no less than 14 times between 1292 and 1341. They needed a large town that could be strongly defended to act as a springboard to service armies destined for incursions across the border. Berwick was too vulnerable a location, frequently changing hands between the Scots and the English, the last time in 1482. Newcastle was always going to be the royal stronghold in the north. It received a long sequence of royal privileges: made a chartered borough, given the honour of having a mayor and, in 1400, made a county in its own right, separate from Northumberland, a status that it retained until 1974 when it became part of Tyne and Wear Metropolitan County. Successive monarchs gave trading and mining privileges, monies and recognition to Newcastle that guaranteed it

OPPOSITE: Gian Tommaso Scala, *View of Newcastle upon Tyne* (1545) [BL] Scala was a peripatetic military engineer, born in the Republic of Venice and known to work in Cyprus, Italy, France, England, Scotland, Flanders and Germany.

becoming a strongly royalist town, as was evident in its loyalties in the Civil War.

Defences therefore played a critical part in the way the town developed. One of the best depictions of Newcastle's defensive walls is a very early panorama. Its date is uncertain but was probably drawn around 1545 by the Italian Gian Tommaso Scala when new fortifications were being constructed in Tynemouth in response to an order from Henry VIII for surveys of defences (*see* 1545a). 'Castell Hill' is shown as a steeply rising prominence from which the castle dominates the town and overlooks the Tyne. The bridge is drawn very effectively with its two gate towers and the chapel at its northern end, while the extensive quay, backed by the defensive wall, is depicted dramatically, and Pilgrim Street is shown leading straight to its northern gate. As one would expect from a military engineer, the defensive walls are very prominent, with their numerous towers and gates, although the artist clearly had difficulty depicting the true line of the wall in the west and the panoramic perspective inevitably obscures part of the northern stretch.

Scala's panorama shows the rather ambiguous relationship between the castle and the town's defensive walls. The castle came first. The distinctive castle mound had been the site not only of a Roman fort but also of an eleventh-century wooden Norman motte and bailey. This was replaced by a stone castle with an impressive keep in 1172–75. The Black Gate was built as an additional barbican to strengthen the defences of the north gate 70 years later. John Dobson's later architectural drawings show that the keep was a formidable building, in parts with walls 15 feet wide, and with a chapel, hall, governor's suite and a well driven down some 100 feet into the rock on which the castle was built.

However, because of the town's topography it is difficult to see how walls and castle could best be combined. It is likely that the first medieval walls were built in the Norman period and, whilst their lines are uncertain, it is probable that their southern line stretched from the castle and followed the rim of the plateau above the river. This meant that the lower town, with its rich merchants and the growing trading significance

Plan de Newcastle ou Neuchastel (*c.*1650) [NCL] The scale of the plan is given in paces suggesting it pre-dates surveying with instruments. The number of towers along the wall is somewhat exaggerated.

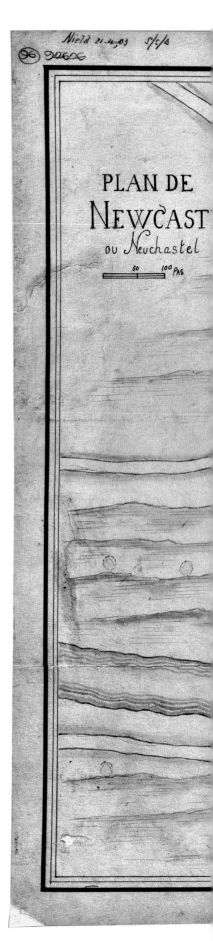

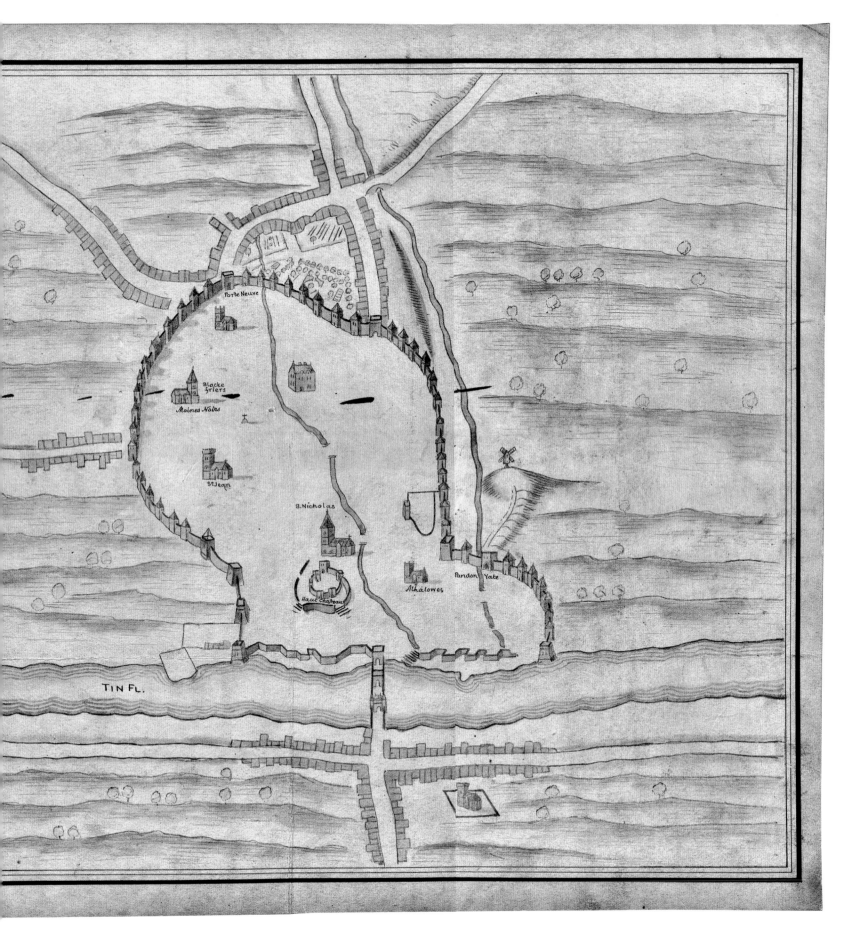

Porte Neuve

Blacke
friers

Moines Noirs

St Jean

S. Nicholas

Haut Chasteau

Alhalowes

Pandon Yate

TIN FL.

CASTLE

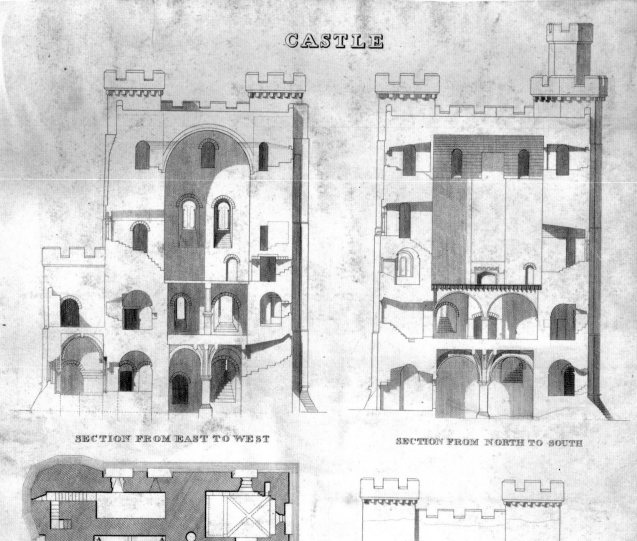

SECTION FROM EAST TO WEST

SECTION FROM NORTH TO SOUTH

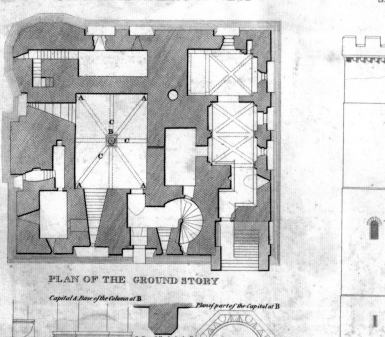

PLAN OF THE GROUND STORY

Capital & Base of the Column at **B**

Plan of part of the Capital at B

Corbel at the Angles A

Section of the Arch ribs C

Plan of part of the Base

Scale of   Feet

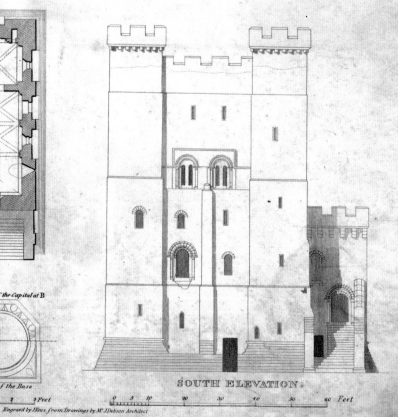

SOUTH ELEVATION

0   5   10   20   30   40   50   60 Feet

Engrav'd by J.Roe, from Drawings by Mr J.Dobson Architect

of the river frontage, was not protected. So, while the castle was then an integral part of the defensive wall geometry, it left large areas of the town without protection. The alternative strategy was to build an encircling wall encompassing the western and eastern areas, leaving the castle standing isolated from the immediate defences. Which is just what had happened by Scala's time.

The French map of *c.*1650 makes the completed walls abundantly clear. It is one of 116 plans of fortifications in England, France, the Netherlands and Germany, 'drawn by a French artist' around 1650. Its purpose was to show defences, so besides the main churches, Blackfriars and a few civic buildings, there is little other detail of the town itself. Instead, the focus is on the castle, as well as some substantial earthworks outside the walls in Shieldfield (used in the Civil War) and the steep-sided valleys of the Lort and Pandon Burns. The walls themselves were clearly a major and imposing feature. They were started in 1265 and progressed slowly until a new war with Scotland hastened their completion on the landward side in 1296, funded both by a local tax, or murage, and by monies from the Crown. When complete, they stretched around the town for two miles and were over six feet wide and up to 25 feet high. They enclosed most of the northern extent of the town, dropped precipitously down to the river on its western side and on its eastern side turned sharply to include Pandon, which had been incorporated into the town in 1299. However, it was not until the early fifteenth century that the walls ran along the frontage of the Tyne. The steepness of the south-western section is well illustrated by the popular title of 'Breakneck Stairs' attached to the 140 that had to be negotiated climbing up the rampart walk (before demolition) from the Close Gate to White Friars Tower. The walls had six principal gates and 17 towers. The gates, clockwise from the Close, were Close Gate, West Gate, New Gate, Pilgrim Gate, Pandon Gate and Sand Gate. The towers originally sported figures of military men, as does the gateway of Alnwick Castle to this day. The walls were supplemented by a ditch, the King's Dyke, which in parts was over 30 feet wide and 15 feet deep.

Kept in good repair, the walls constituted a formidable obstacle. They withstood a siege by the Scottish King David II in 1342 and repulsed an invasion by the Scots in 1388. However, in the Civil War a Scottish army laid siege to the town for three months and breached the wall by using mines on the stretch between Close Gate and White Friars Tower.

Much of the wall was demolished after the Act of Union between England and Scotland in 1707. The first section to go, in 1763, was the stretch along the Quayside which had become a serious impediment to the export trade. The Georgian reconstruction of the town entailed the progressive demolition of most of the northern and eastern sections of the wall. As a result, today there remain only short stretches, largely in the west behind Stowell Street and Hanover Square but also some isolated sections in the east. The gates, which had become a major hindrance to traffic, were progressively demolished between 1795, when Pandon Gate was removed, and 1823, when New Gate was demolished.

Even before 1707, mainly due to the completion of the walls, the castle had lost much of its rationale and became dilapidated. The keep was used as a prison between the sixteenth and eighteenth centuries, and houses, shops and inns were built in the Castle Garth. So disregarded was it by the mid nineteenth century that the railway line to Edinburgh was driven between the Keep and the Black Gate.

OPPOSITE: John Dobson, *Castle Keep: Sections and Elevations* (*c.*1847) [ROB] Dobson was centrally involved in the restoration of the Castle Keep on behalf of the Society of Antiquaries in 1847.

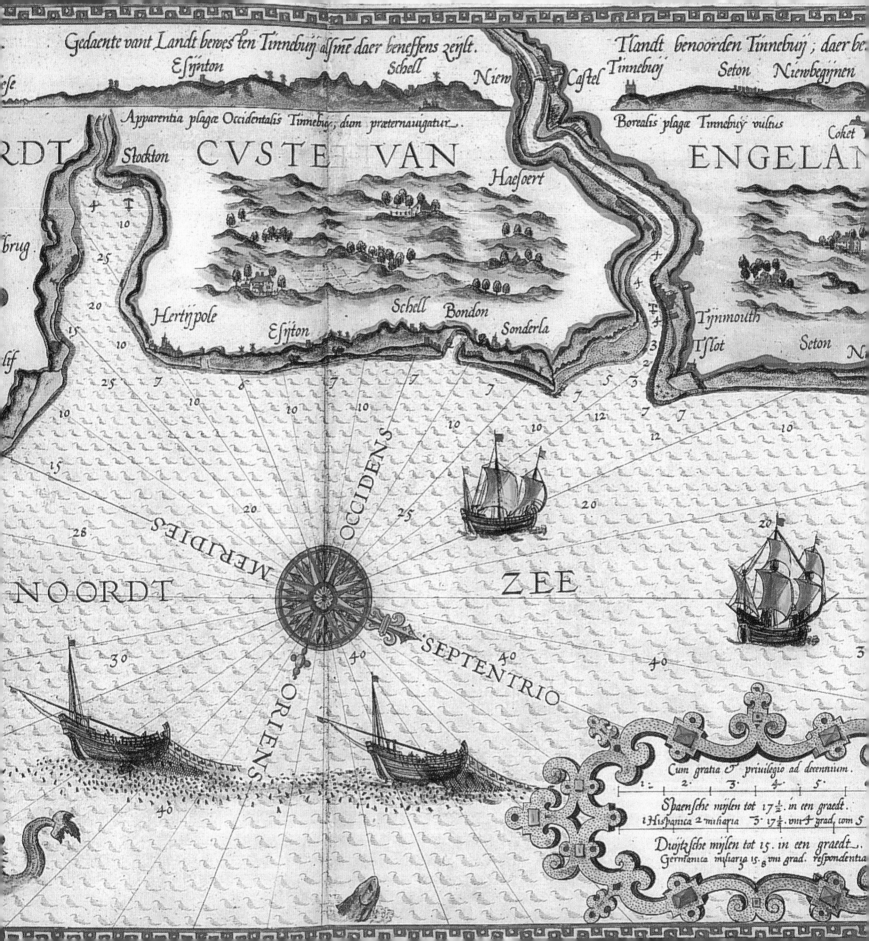

Gedaente vant Landt bewes ten Tinnebuÿ alsme daer beneffens zeÿlt.

Tlandt benoorden Tinnebuÿ; daer be

Esÿnton   Schell   Niew   Caſtel   Tinnebuÿ   Seton   Niewbegÿnen

Apparentia plagæ Occidentalis Tinnebuÿ, dum præternauigatur.

Borealis plagæ Tinnebuÿ vultus

Coket

RDT   Stockton   ENGELAN

CVSTE VAN

Haeſoert

brug.

Hertÿpole   Schell   Bondon   Tÿnmouth

Eſÿton   Sonderla   Tſlot

lif   Seton   N

OCCIDENS

MERIDIES

NOORDT   ZEE

ORIENS

SEPTENTRIO

Cum gratia et priuilegio ad decennium.

Spaenſche mÿlen tot 17½. in een graedt.
1 Hiſpanica 2 miliaria 3. 17½. viii 4 grad. tom 5

Duÿtſche mÿlen tot 15. in een graedt.
Germanica miliaria 15. 8 viii grad. reſpondentia

1585

The perils of the unimproved Tyne

Notwithstanding the vast . . . income earned by the town of Newcastle over hundreds of years of trading, very little was spent on maintaining the navigability of the river itself.

P.D. Wright, *Life on the Tyne*, 2014

Before the sixteenth century British sailors navigated mainly from experience or using information passed on orally, but with the establishment of the Corporation of Trinity House in 1514 as a royal charity, the training of pilots and methods of navigation in inshore waters improved considerably. Trinity House, with its London base, had special responsibility for lighthouses and other navigational aids. The provision of navigational charts was part of this progress, and the first known such chart specifically for the Tyne is attributed to a mariner, Richard Poulter, probably about 1590.

But far more sophisticated was the chart of the north-east coast from the second volume of Lucas Janszoon Waghenaer's groundbreaking *Spieghel der zeevaerdt* of 1585 (translated as *Mariner's mirror*), the first printed sea atlas. It covers the coast between Robin Hood's Bay and 'Coket [Coquet] Island'. Waghenaer played a major role in the early development of Dutch cartography, with his maritime charts making him a seminal figure in the golden age of Dutch cartography. His map has all the hallmarks of that period: confident engraving (by Jan van Doetichum); the title and scale bar set with bold strapwork; large areas of sea filled with fish, sea monsters and fighting ships; and, unusually, two fishing trawlers dragging their nets. That this was a chart aimed at assisting navigation is clear not just from the elaborate compass rose from which radiate no fewer than 32 compass lines, but from the sea depths (shown in fathoms) and the panorama of the coast as seen

OPPOSITE: Detail of Lucas Waghenaer, *Cueste der noordtcuste van Engelandt* (1585) [AUTH]
Presented from the seaman's perspective. West is at the top.

from the sea. These aimed to assist the seafarer at a time – before longitude could be accurately measured – when dead reckoning based on speed and direction was the only way in which navigators could estimate their current position.

Waghenaer had been a ship's chief officer, but later turned his hand to marine cartography. He produced the first volume of the *Spieghel der zeevaerdt* in 1584. It was both innovative and a huge market success with numerous reprintings. His depiction of the north-east coast is somewhat generalised, although most of the named towns can readily be identified: from the south on this extract, Hartlepool, Easington, possibly Seaham, possibly Boldon, Sunderland, Tynemouth, Seaton and Newbiggin. The Tynemouth fort is indicated both on the chart and in the panorama, and a number of settlements – North Shields, Jarrow and Wallsend – are shown as clusters of buildings hugging the riverbanks. Gateshead and Newcastle are depicted as clearly the largest settlements and linked by the Tyne Bridge.

The water depths are the most telling element. There is a striking contrast between the depths at the mouths of the Tyne and Tees. The latter is between ten and 25 fathoms as far as Stockton where the depth lessens to four fathoms. In contrast, the Tyne is shown as just two fathoms at the bar across its mouth and nowhere more than four fathoms up to Jarrow. The challenges of entering and then navigating the Tyne are obvious.

By the end of the seventeenth century, there had been only minimal improvement to the Tyne. Greenville Collins' chart shows a river that would have been familiar to seafarers from Waghenaer's time. Collins had been a ship's officer in the Royal Navy, doubtless fully aware of the imperfections of existing nautical charts. He gained royal support to survey the British coasts and then spent almost a decade undertaking what was the first coastal survey of Britain by an Englishman. His *Great Britain's Coasting Pilot* was published in 1693, and subsequently went through numerous re-issues in the eighteenth and nineteenth centuries.

The map is dedicated to the 'Master and Bretheren' of Trinity House at Newcastle. Collins himself became an 'elder

brother' of Trinity House. His chart focuses on the Tyne and extends north along the coast to Seaton Sluice and Blyth. It incorporates inset details of the mouths of the Wear to the south and the Blyth to the north. Like Waghenaer, it has a compass rose and ships in the sea. Its main interest, however, is the detail of the Tyne which is displayed upstream to Newcastle. The navigational challenges of the river are again well shown. The bar at the river mouth has a clearance of only eleven feet. To its south the Herd Sands are shown, and to the north Sparrow Hawk, and the rocky shallows of the Black Middens are indicated along with the warning presence of the major lighthouse at Tynemouth, built in 1665. Given the difficulties presented by the river mouth, the appearance of the Low and High Lights in North Shields is not surprising, with Collins illustrating the approach to the river guided by aligning the two lights.

Collins also shows and names the series of towns, villages and quays along the river. Many of the principal churches are featured – St Mary's in Gateshead, St Ann's outside Newcastle's walls, St Bede's in Jarrow – and, within the walls of Newcastle, the castle and St Nicholas Church are depicted. There is also evidence of the growing industrial activity of the time: glass works to the south of the Ouseburn, salt pans in South and North Shields, and coal pits along the coast towards Blyth.

Even into the eighteenth century, there had still been relatively little change to the Tyne, as is evident from the chart by Gerard van Keulen, which comes from the third volume of his *Zee-Fakkel*. The van Keulen family firm was in business for over 200 years at Amsterdam, making charts, navigational instruments, maritime atlases and pilot guides. It was founded by Johannes van Keulen who published two enormously successful atlases, the *Zee-Atlas* and the *Zee-Fakkel*, which initially appeared in the 1680s and continued to be published in different languages and with additions and changes well into

OPPOSITE: Greenville Collins, *To the worshipful Master & the rest of the bretheren of Trinity Hose, Newcastle upon Tyne* (1693) [AUTH] North Sea coast from Sunderland to Blyth from Greenville Collins' famous *Great Britain's Coasting Pilot, 1693.*

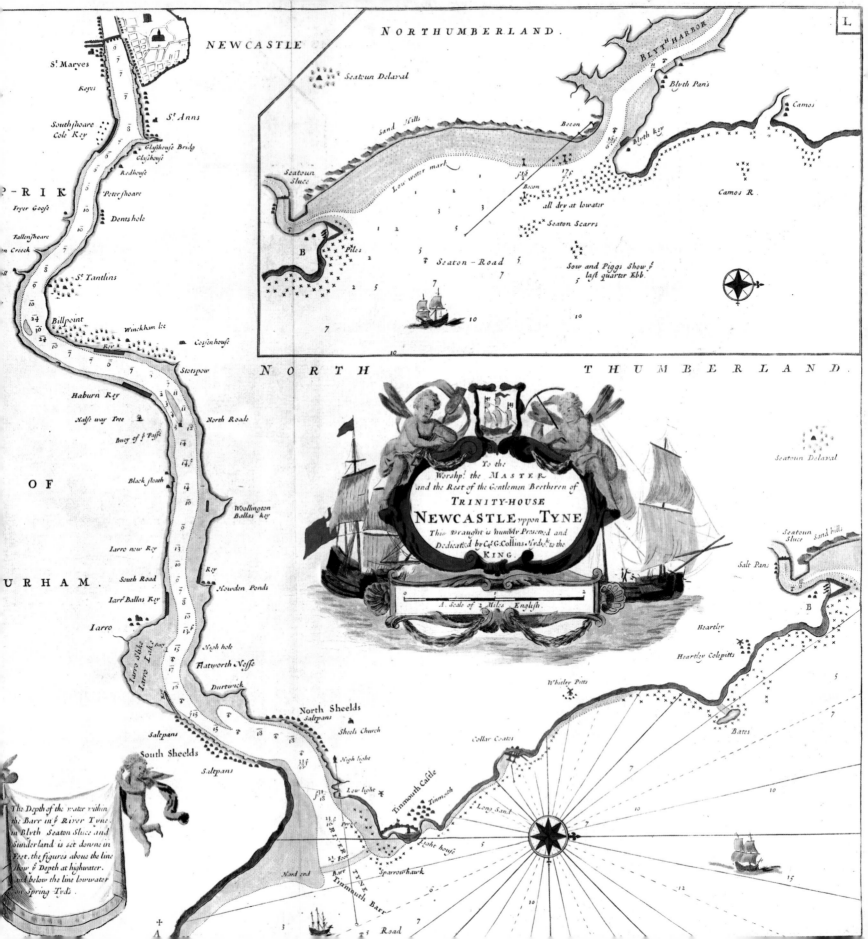

L

NEWCASTLE

BLYTH HARBOR

St. Maryes
Keyes
St. Anns
Southshoare Cole Key
Glasshouse Bridg
Glasshouse
Redhouse
Peter shoare
Dents hole

P-RIK
Fryer Goose
Fallenshoare
Creek
St. Tantlins
Billpoint
Winckham lee
Key
Cossenhouse
Stotspow

Haburn Key
Halfe way Tree
Bucy of y Passe
North Roade

Black steath
Woollington
Ballas key

OF
DURHAM.
Iarro new Key
Key
South Road
Howdon Ponds
Iars Ballas Key
Iarro
Iarro Slike
Iarro Lake
High hole
Flatworth Nesse
Durtwick

Salipans
South Sheelds
Salipans

North Sheelds
Salipans
Sheels Church
Nigh light
Low light
Tinmouth Castle
Tinmouth

Seatoun Delaval

Sand Hills
Becon
Blyth Pans
Camos
Blyth key
Low water mark
Becon
Camos R.
all dry at lowater
Seaton Scarrs
Seaton — Road
Sow and Piggs Show y
last quarter Ebb.

Piles
B

To the
Worshp! the MASTER
and the Rest of the Gentlemen Breetheren of
TRINITY-HOUSE
NEWCASTLE uppon TYNE
This Draught is humbly Presented and
Dedicated by Capt G. Collins, Hydrogr to the
KING.

A Scale of 2 Miles English

Seatoun Delaval

Seatoun Sluce
sand hills
Salt Pans
B
Heartley
Heartley Colepitts
Wheeley Pitts
Bates
Collar Coates

Long Sand

River Tyne
Prict
Light house
Sparrowhawk
Hard end
Tynmouth Barr
Road
A

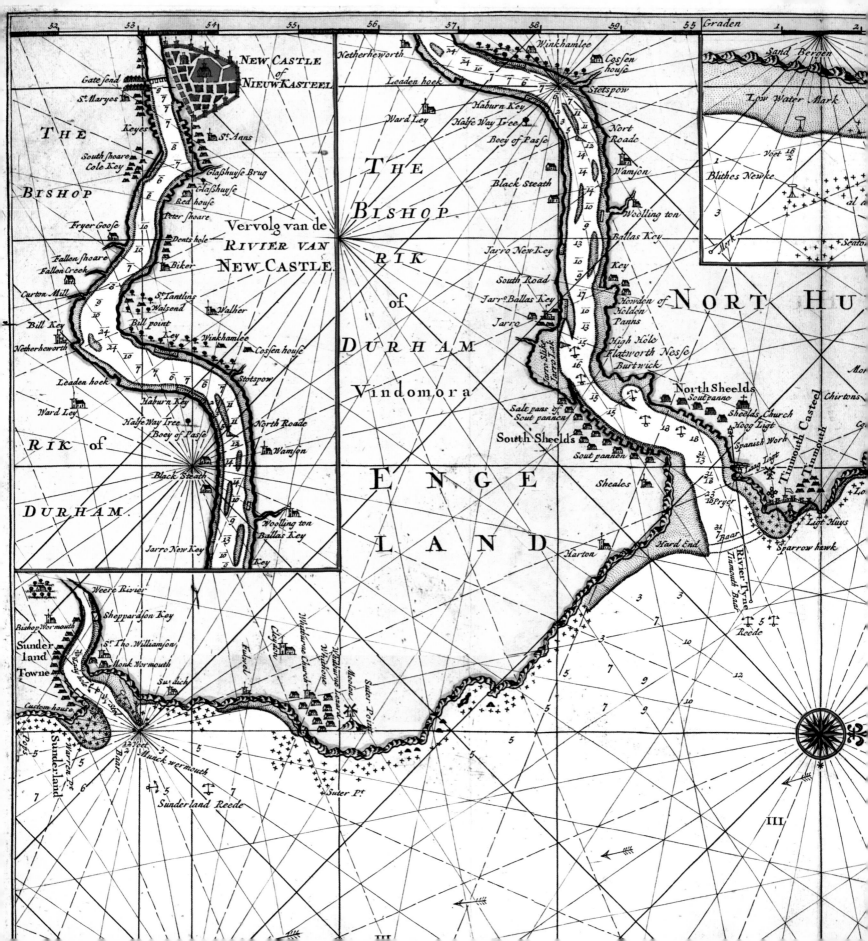

Graden

NEW CASTLE of NIEUW KASTEEL

THE BISHOP RIK of DURHAM

Gate send
St. Maryes
Keyes
South shoare
Cole Key
Fryer Goose
Fallen shoare
Fallen Creek
Curton Mill
Bill Key
Netherheworth
Leaden hoek
Ward Ley
St. Anns
Glashuyse Brug
Glashuyse
Red house
Peter shoare
Dents hole
Biker
St. Tantlins
Walsend
Bill point
Key
Winkhamlee
Cossen house
Stotspow
Haburn Key
Halfe Way Tree
Boey of Passe
North Roade
Wamson
Black Steath
Woolling ton
Balla's Key
Jarro New Key
Key
Walher

Vervolg van de RIVIER VAN NEW CASTLE.

THE BISHOP RIK of DURHAM Vindomora

Netherheworth
Winkhamlee
Cossen house
Leaden hoek
Stetspow
Haburn Key
Halfe Way Tree
Boey of Passe
Nort Roade
Black Steath
Wamson
Woolling ton
Balla's Key
Jarro New Key
Key
South Road
Jarro Ballas Key
Jarro
Howden of Holden Panns
High Hole
Flatworth Nesse
Burtwick
North Sheelds
Sout panne
Sheelds Church
Hoog Ligt
Spanish Worh
Salt pans of Sout pannen
South Sheelds
Sout pannen
Sheales
Marton
Hard End
Pryer
Baar

NORT HU

Sand Bergen
Low Water Mark
Voet
Blithes Newke
Mork
Seaton

ENGE LAND

Tinmouth Casteel
Tinmouth
Lang Ligt
Ligt Huys
Sparrow hawk
Rivier Tyne Tinmouth Baar
Reede
Chirtons

Weere Rivier
Sheppardson Key
Bishop Wormouth
Sunderland Towne
St. Tho. Williamson
Monk Wormouth
Su dich
Cley don
Fulwel
Whitburne Church
Whitborne
Morton Zeland
Suter Point
Custom house
Sunderland
Warren Pt Baar
Munck wermouth
Suter Pt
Sunderland Reede

III

the eighteenth century. The *Zee-Atlas* appeared in no fewer than nine editions in its first five years, with charts drawn by Claes Janz Vooght. The *Zee-Fakkel* (which in English would read '*Sea Torch*') was first published in five volumes between 1681 and 1684. By 1704, Johannes had retired, and his son Gerard took over the business until his death in 1727. Gerard was a talented engraver and a mathematician, and his original chart reflects these attributes, incorporating an elaborate series of compass roses – eight in the sea and a further four within the three inset plans.

The extract from van Keulen's chart shown here indicates how little had changed since Waghenaer's map. The difficult entry to the Tyne still presented severe hazards to shipping. Van Keulen shows the bar as now having a mere seven feet of water. The Herd Sands and Black Middens constricted the river mouth and the passage to the river was clearly dangerous despite the navigational aids of the High and Low Lights and of the 'Ligt Huys' above Sparrow Hawk within the grounds of Tynemouth's Priory and Castle. What he does show more clearly than Collins is the number of islands along the course of the river, adding yet further danger to ships sailing up to Newcastle and to the various sites from which coal was loaded. Newcastle itself is displayed with its walls running continuously around the town except along the river frontage. North and South Shields still have a multitude of salt pans. Oddly, despite all its detail and accuracy, van Keulen did not get the inland geography quite right: his plan suggests that Scotland started somewhere around Seaton Sluice, less than five miles north of Tynemouth.

Gerard van Keulen, *De rivier Tyne of New Castle geleegen aan de oost Kust van Engeland in de Noord Zee* (1730) [AUTH] Note the coal-based development along the Tyne, and the harbour and staithes at 'Collar Coates' serving 'Whitley Pitts'.

c.1595

The 'Coaly Tyne': coal mining and the export trade

Our keelmen brave, with laden keels,
Go sailing down in line,
And with them load the fleet at Shields
That sails from coaly Tyne.

> W. & T. Fordyce, *The Tyne Songster*, 1840

The first reference to the coal trade in the Tyne was in 1367 when 676 chaldrons (one chaldron being approximately 0.9 of a ton) of coal were purchased from Winlaton for Edward III's household at Windsor Castle. The Tyne and its proximity to coal seams were the two elements that initially gave the North East such a commanding role in the mining and export of coal before 1800 when water transport was the only realistic means of moving bulky materials. The glories of the Tyne were celebrated in numerous local songs and verse, but the river nevertheless presented difficulties to navigation: the very shallow depths at its mouth; the numerous shallows along its length; the fluctuations in depth resulting from its tidal reach as far west as Wylam; and Newcastle's low medieval bridge which prevented larger vessels from sailing further upstream. However, the verse from one of the many North East songs captures the essence of the system developed to enable exports to flourish. Coal was moved from staithes on the riverbanks, using smaller three- or four-man vessels known as keels to reach larger vessels anchored at the mouth of the river. The keelmen formed a distinctive labour organisation. In the early eighteenth century it was calculated that around 1,600 keelmen were employed, about 400 being Scots who moved back and forth according to the season, winter being a relatively slack time in the coal trade due to inclement weather for both the

---

OPPOSITE: Detail from *An ancient map of Tynemouthshire and part of Northumberland (c.1595)* [N.EST]
Like many early maps, places are represented pictorially by a church steeple or tower and there is foreshortening of distances, especially east–west. Significantly, staithes, quays and pits are depicted in detail.

J.T.W. Bell, detail from *Plan of part of the Newcastle Coal District* (1847) [ROB] Between 1842 and 1861 Bell produced
six superb huge maps of the Northumberland and Durham Coalfield at a scale of 2½ inches to one mile.

production of coal and its transport. By the end of that century there were 996 keels and almost 4,000 keelmen working on the river. The shortage of wood in the mid sixteenth century acted as a catalyst for the use of coal for domestic purposes as well as in industry, with the east coast, and especially London, providing a growing market. Coal shipments from the Tyne grew to around 250,000 tons in 1600, reached 650,000 tons by 1700 and over a million tons by the end of the eighteenth century.

The initial concentration of coal mining for export by sea was in a small number of geographically limited areas: between Whickham and the river Derwent near Blaydon; and, as shown on the map, at Elswick; the Ouseburn area east of Newcastle; Felling Shore east of Gateshead; and a site west of North Shields. Although this map is undated, it was probably produced a few years after 1590. The map reinforces the fact that the initial exploitation of coal was as near to water as a means of transport as possible: the faint lines linking coal pits with riverside staithes are 'wainways' rather than waggonways, as the latter did not develop until well into the seventeenth century. Wains were four-wheeled carts usually pulled by two oxen and two horses along an agreed routeway.

An intriguing naval skirmish is taking place off the mouth of the Tyne. This has been interpreted as a conflict between a

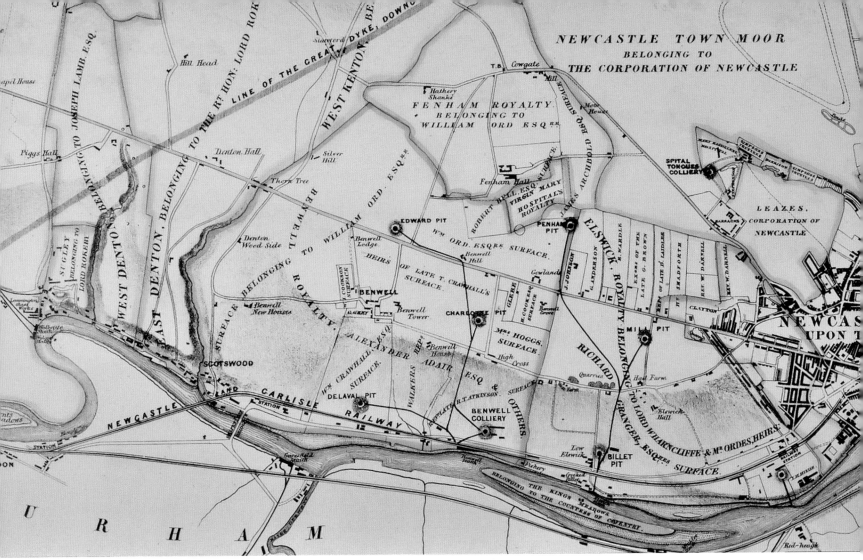

Bell (1847) [ROB] Bell's maps show the ownership of land and associated mineral royalties,
coal pits, railways, roads, settlement, parks and country houses.

naval vessel and some remnants of the Spanish Armada, but possibly the depiction of this confrontation was also intended as a reminder of the perils associated with the east coast export trade. In the early seventeenth century there was mounting concern with the impact of privateers operating out of Dunkirk and Ostend. Such fears were exacerbated with the outbreak of war with France and Spain in 1625, leading to the recommendation that colliers should travel in convoys.

In the early seventeenth century, new pits were opened, especially on the north side of the Tyne upstream from Newcastle at Newburn, Denton, Elswick and Benwell, but always close to the river with a distance of around two miles

from pit to staithe being considered the maximum. This pattern was generally determined by geology, with the coal measures dipping to the east and lying too deep nearer the coast for the technical means of exploitation at this early period. Some coal-owners, aware of diminishing resources in the active pits and conscious of substantial reserves further to the west, began to purchase leases in the north-west Durham area leading to the emergence in 1726 of the so-called 'Grand Allies', a powerful group of coal-owning families who dominated the trade.

Increasing use of waggonways also allowed exploitation further afield, with pits up to ten miles from the river, but during the eighteenth century the pendulum for new exploita-

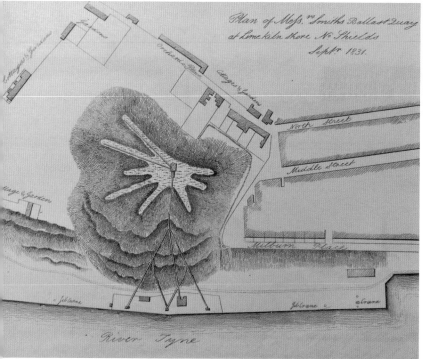

Plan of Messrs Smiths Ballast Quay at Lime Kiln Shore, North Shields
(1831) [SANT] The plan shows the mechanism used to haul ballast to
the top of the mound and then distribute it.

tion swung to the much deeper eastern measures, especially
the High Main Seam at Wallsend, Hebburn and Jarrow. The
sinking of these pits was enabled by innovations in ventilation
and drainage, permitting mining to take place at greater depths
where water ingress had been a major problem as well as in
larger pits employing up to 400 men. Central to this was the
introduction of Newcomen steam pumps. In 1714 there were
only four such engines nationally, two of them on Tyneside
(*see* 1749).

The area to the east of the Tyne Bridge was dominant by
the beginning of the nineteenth century and was characterised
by in-migration from the western coalfield. Within a few
decades, new areas were being exploited further north beyond
the Ninety Fathom Fault. George Stephenson's moves from
Wylam to Willington Quay and thence to Killingworth
exemplify the chronological pattern of coalfield exploitation

and labour migration. The type of coal mined in the Tyneside
area had been predominantly household coal, with poorer
quality 'small coals' used in industries such as salt and pottery.
But the demand for steam coal arising from the vast scale of
mechanisation in the nineteenth century prompted a shift to
deeper reserves at greater distances from the river. In particular,
the Northumberland coalfield produced steam coal, and its
development created a complex pattern of collieries, railways
and staithes. The invention of steam locomotives and steam-
driven marine vessels both stimulated and enabled further
exploitation of this wider area.

This expansion was reflected in the cartographic history of
Newcastle and Tyneside, and the emergence of a specific genre
of mapping related to coal exploitation. The complex logistics
of access, granting of royalties and wayleaves, and the transport
of coal to a suitable location for export led to the production
of a large number of detailed local maps akin to rural estate
maps. The term 'Great Northern Coalfield' now began to be
used, as reflected in the production of a remarkable series of
larger-scale maps of the coal industry, including T. Wilson's
Plan of the River Tyne (1754), John Gibson's *Plan of the
Collieries on the Rivers Tyne and Wear* (1787), William
Casson's *Plan of the Rivers Tyne and Wear* (1801), D.
Akenhead's *The Picture of Newcastle upon Tyne* (1807) and
T.Y. Hall's *Map of the Great Northern Coalfield* (1854). Excellent though most of these maps are, they were eclipsed by the
publication, between 1842 and 1861, of J.T.W. Bell's magnificent series of maps of the coalfield. This consisted of six large
maps at the detailed scale of 2½ inches to the mile, each
showing land ownership, mineral royalties, and the location
of collieries and railways linked to named riverside staithes.

Long before the huge expansion of the mid and later
nineteenth century, the growth in exports posed the challenge
of disposing of the ballast brought in by the colliers. In 1549
the burgesses of Newcastle extended the town boundary to
incorporate part of Byker in order to extend the quayside and
have more space to deposit ballast. In 1702–03, of the 2,280
coastal vessels arriving in the river, 2,074 carried ballast which
had to be jettisoned before they could load with coal. This had

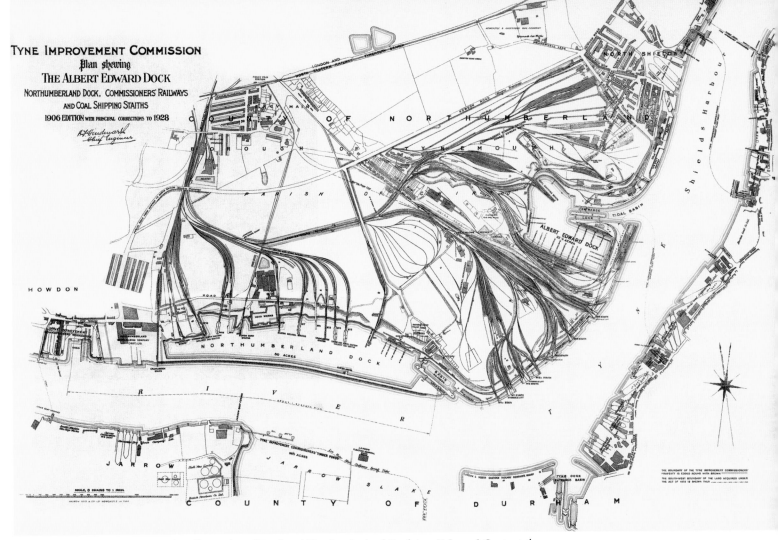

Tyne Improvement Commission, *The Albert Edward Dock and Northumberland Dock* (1928) [AUTH] Contrast the
coal exporting infrastructure shown here with the keelboat-based system shown in the Bucks' panorama of 1745.

to be discharged – at a cost – at official sites which became a profitable and prominent topographical feature. Thomas Oliver noted in 1830 that within the space of just 11 hours 300 tons of ballast could be taken from the ships and deposited at St Anthony's Ballast Quay. Around 30,000 tons were added to this artificial mound annually. At Wincomblee Ballast Quay about 25,000 tons were added annually with 18 pence being paid for every ton. To avoid the charge, some ship's masters simply dumped ballast in the river. Local records of fines attest to the frequency of this occurrence. But the ballast hills themselves sometimes became too large and erosion caused material to be deposited into the river, adding to the navigational problems.

In the later nineteenth century, much improved techniques of sinking the pits and the further development of railway connections enabled deep mining in locations more distant from Tyneside, including submarine exploitation especially in the hidden coalfield of east Durham. Output from the entire North East Coalfield (including Durham) peaked at 56 million tons in 1911. Linkage to the Tyne remained important, as the immense infrastructural developments of Northumberland Dock and Albert Edward Dock illustrate. It is slightly ironic that the town of Ashington is popularly most associated with North East coal mining, but its pits were actually a relatively modern creation dating from just a few decades before the First World War.

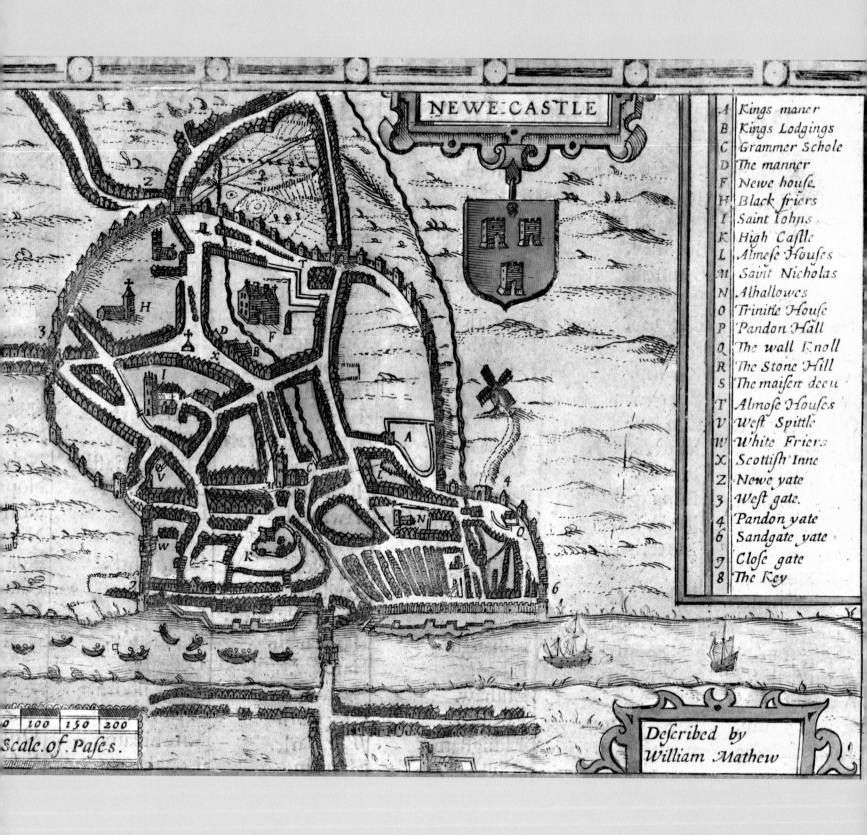

NEWE:CASTLE

A	Kings maner
B	Kings Lodgings
C	Grammer Schole
D	The manner
F	Newe house
H	Black friers
I	Saint Iohns
K	High Castle
L	Almese Houses
M	Saint Nicholas
N	Alhallowes
O	Trinitie House
P	Pandon Hall
Q	The wall Knoll
R	The Stone Hill
S	The maisen deeu
T	Almose Houses
V	West Spittle
W	White Friers
X	Scottish Inne
Z	Newe yate
3	West gate
4	Pandon yate
6	Sandgate yate
7	Close gate
8	The Key

scale.of.Pases.

100   150   200

Described by
William Mathew

1610

Speed's enduring town plan

Many layers and levels and heights and materials are woven into a rich fabric: but Newcastle's medieval origins can still be found at its heart.

R. Fawcett, 'The Architecture of Tynemouth Priory Church',
in J. Ashbee and J. Luxford, *Newcastle and Northumberland:
Roman and Medieval Architecture and Art*, 2013

Compared to towns in Continental Europe, maps of British towns were few and far between until the seventeenth century. Many of those that were produced were done to plot town defences. The first real cartographic coverage of British towns came with John Speed's *Theatre of the Empire of Great Britain* of 1611 which was published alongside his separate *History of Great Britain*. Together they aimed to celebrate the 'countries glory', an important element of which was the growth of towns under the Tudors. This prompted Speed to include on his county maps insets of the 'most principall townes cyties and castles . . . in such estate and forme as at this day they are'. In total, there are 73 town plans, including 16 Welsh towns, four Irish towns and the single Scottish example of Edinburgh. While his county maps were almost entirely copied from other sources – not least Christopher Saxton's superb maps in his groundbreaking atlas of 1579 – Speed produced most of the town plans himself. In his preface, he says that 'some have been performed by others, without Scale annexed, the rest by mine owne travels, and unto them for distinction sake, the Scale of Paces . . . five foote to a pace I have set . . .' It is clear, for example, that he drew on William Cunningham's 1559 woodcut plan of Norwich and William Smith's plans of Bristol, Westminster, London and Chichester, as well as on Richard Lyne's 1574 plan of Cambridge which was the first town plan engraved on copper by an English cartographer.

OPPOSITE: John Speed, *Newe: Castle* (1610) [AUTH] Note 'Described by William Mathew',
bottom right, suggesting that Speed may not have been the principal author.

31

However, he almost certainly 'surveyed' at least 50 of the towns, probably pacing distances and possibly using a plane table for angular measurements.

The Newcastle plan is especially intriguing since it is the only one of Speed's plans that adds an attribution to its source, saying 'Described by William Mathew'. It is uncertain who Mathew was, although John Brand in his 1789 *History* of the city notes the name in connection with an inquisition into the condition of the old castle in about 1610, and Mathew has been described as an almanac-maker. In 1609 he surveyed a 'platt' of land in the small village of Nunstainton (near Ferryhill in County Durham) for the Dean and Chapter of Durham Cathedral, so he had some mapping expertise, although there is no record of other maps or plans that he produced. A further puzzle is that, while Speed suggests that the plan was 'described' by Mathew, he shows the scale bar with 'A scale of pases', which is generally taken as signifying that the plan was Speed's own work, yet Speed's travels from town to town between 1606 and 1608 do not seem to have included the North East. Of the three North East towns included in the *Theatre*, there were existing plans of Durham and Berwick on which he could have drawn, but there appear to be no appropriate maps of Newcastle that he could have copied.

Speed's plan, surveyed in 1610, paints a clear picture of Newcastle as it was in the early seventeenth century. It is shown as completely walled. Five of the gates are named. The old bridge is pictured with its towers and houses. The four existing churches – St John, St Andrew (shown but not named), All Hallows (All Saints) and St Nicholas (which became the cathedral in 1882) – are each depicted by conventional pictographs, as is St Mary's in Gateshead. The river is shown with numerous vessels, rowing boats upstream of the bridge and masted ships downstream. The grammar school is shown in its early location in the grounds of St Nicholas, and 'West Spital' (the hospital of Mary the Virgin to which the school later removed) is noted on Westgate Street. The White Cross, mentioned as early as 1410, is shown at the junction of Newgate Street and Low Friar Street. The King's 'Maner', which Henry had taken over from the Austin friars, is named in the south-east quadrant of

the town. Speed shows the 'Newe House' built on the site of St Bartholomew's Nunnery, giving a clear impression of the size of the grounds on which the nunnery had stood. Two streams are shown: to the east is the line of Pandon Burn lying just outside the town walls; and through the centre of the town the Lort Burn cascades down to join the Tyne close to the old bridge. The Lort was the stream that was progressively filled in as the town's early development progressed and Speed's plan is one of the few that show it (*see also* 1545b).

Harold Whitaker, in his *Descriptive List of the Maps of Northumberland*, listed over 20 reprints of Speed's *Theatre*, six of these being in the eighteenth century alone. The county map of Northumberland had some minor changes in these various editions, but the inset plan of Newcastle remained unchanged. So, for not far short of two centuries it was as though the town stood still. And this was reinforced by the frequency with which other cartographers published virtually identical copies of Speed's plan under their own names: for example, Hermannides in his 1661 atlas of town plans, *Britannia Magna*; Ralph Gardner's *England's Grievance Discovered* in 1655; an inset on Philip Lea's 1689 version of Saxton's county map, which itself was reprinted four times up to 1770; J.C. Beer in his *Gross Britannien* of 1690; and Pieter van der Aa in his massive 27-volume *La Galerie Agreable du Monde*, published in 1729.

While it seems astonishing that unchanged plans could continue to attract custom over this long period, there was in fact relatively little change to the townscape of Newcastle until the second half of the eighteenth century. Corbridge's 1723 surveyed plan (*see* 1723) provides much more detail and accuracy than does Speed, but it portrays a town essentially little different from that of the early seventeenth century. Similarly, the prolific cartographer, Isaac Thompson (1704–76), produced an accurately surveyed plan in 1746. Thompson's patron was the Duke of Cumberland, and the plan was prompted by the 1745 rebellion when all gates apart from Newgate, Sandgate and the gate on the Tyne bridge were walled up and cannon set upon the town walls. Some of Thompson's features reflect military considerations; for

Isaac Thompson, *Plan of Newcastle upon Tyne* (1746) [BL] Thompson (1704–76) was a Quaker land agent and surveyor who produced many estate plans. In 1739 he became owner of the *Newcastle Journal*, rival to the well-established *Newcastle Courant*.

example, the details of gates and towers and the distances between them. Thompson also attempted to represent the local topography with shading of the cliff edge behind the Quayside area, as well as along the steep-sided valleys of the Pandon and Lort burns. All public buildings are identified, along with the details of burgage plots, gardens and orchards. Yet his fine plan confirms the stability of Newcastle's urban form and has the same basic outline as Speed's, with only a limited amount of new development appearing along the riverbanks in the east around Pandon, St Ann's Church and the Keelmen's Hospital.

Speed's plans are doubly important since not only did he provide the first ever plans of many towns but also he largely avoided perspective views, choosing instead to produce vertical ground plots. His tiny sketches of houses and other buildings were set on what were essentially vertical plots of the towns in which the scale was (ostensibly) the same throughout the plan. In the longer term, such linear ground plans were to become the cartographic norm, so Speed's inserts on his county maps broke new ground for English cartography by providing a comprehensive array of English town plans. He helped to create an interest and demand for town plans which had not existed previously. Certainly, his Newcastle plan set the marker for our perception of the town's form over the long period from the Stuart to the Hanoverian eras.

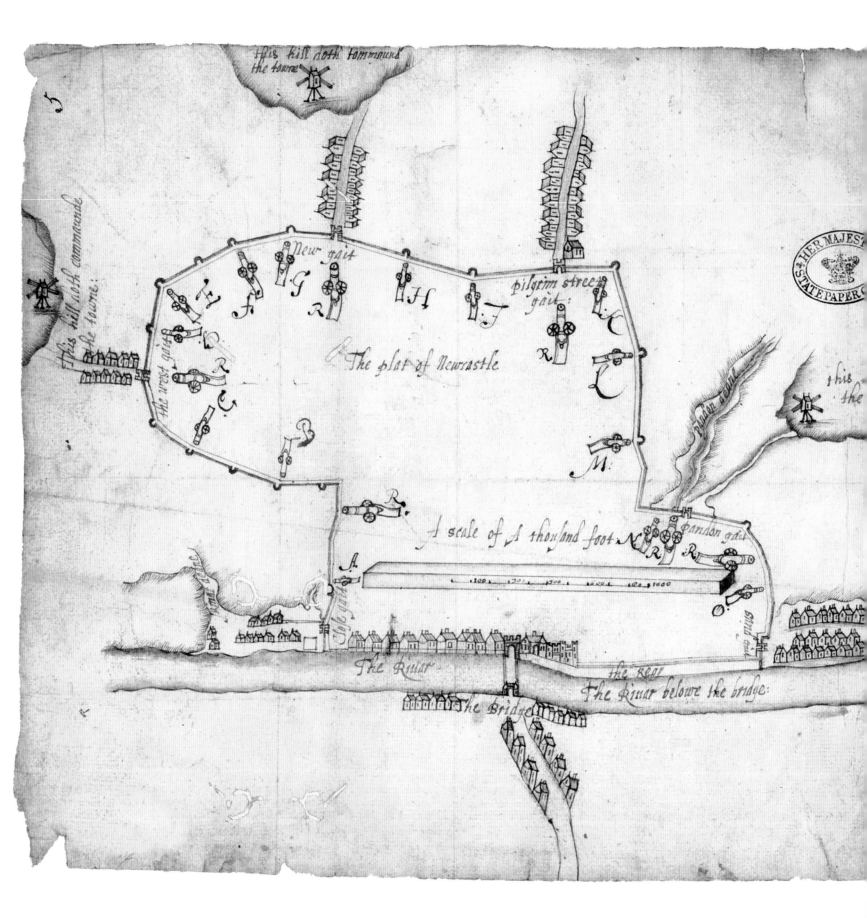

This hill doth tommaund the towne

This hill doth commaunde the towne:

New gait

E
F
G
R
D
R
L
B

pilgrim street gait:

K
C
C

H
J

R

Pladen deina

this the

M

The plat of Newcastle

The west gait

R

A scale of A thousand foot

100 300 500 1000 1600

N R R

pandon gait

A

The gate

sand gait

The Riuar

the Reat

The Riuar belowe the bridge:

The Bridge

1639

Fortiter defendit: *the impact of the Civil War*

19 October 1644, a grey day of autumn, smoke eddied over the scorched walls of the northern city and the jumble of close-packed buildings gapped and scarred by shell. Civil strife . . . was now enjoying a full, bitter flowering. In part this was an ancient grudge.

R. Serdiville and J. Sadler,
The Great Siege of Newcastle, 1644, 2011

Newcastle was at the heart of many of the domestic wars of the seventeenth century and its Royalist sympathies were openly celebrated by William Gray, its first local historian, in his *Chorographia* (1649). The so-called 'Bishops' Wars' that preceded the Civil War focused on the town as a strategic prize and, in the Civil War itself, Newcastle valiantly held out against a long siege by Covenanters but was eventually occupied. After the capitulation of the Royalist forces, Charles I was kept in Newcastle before being taken to London to be put on trial.

The Bishops' Wars were prompted by Charles I's attempt to impose an episcopal system on the Church of Scotland against the wishes of the 1638 National Covenant in Scotland which aimed to expel bishops from the kirk. The Covenanters' army, under the experienced leadership of Alexander Leslie (later Earl of Leven), moved towards Berwick in 1639 and defeated a significantly less professional English army based in the town. Charles avoided a formal battle by meekly agreeing to let Scotland settle the dispute over the governance of the kirk. Meantime, he ordered his new military commander of Newcastle, Sir Jacob Astley, to strengthen the defences of the town. The map opposite is one of two sketches which Astley had prepared for this purpose.

In the following year, the Covenanters' army marched into Northumberland and the English army retreated before it. The

OPPOSITE: *The 'plot' of Newcastle: Sketch plan of defences* (1639) [NA] Some 20 gun emplacements are plotted on the map with the weakest defended area being in the south-west. This is where the walls were finally breached in the siege of 1644.

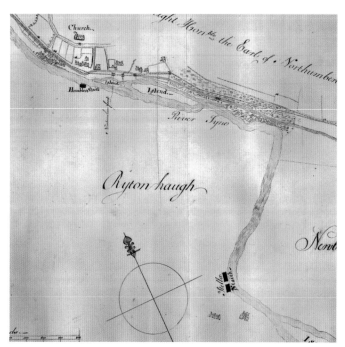

Ryton Haughs: Site of the Battle of Newburn Ford (1763) [GLH] Stella Hall is the building drawn on the bottom right of the map, below Stella Staiths.

garrison in Newcastle had expected the Scottish army to attack the town from the north, but instead it skirted around the west of the town looking to cross the River Tyne at Newburn, the first fordable point above Newcastle. The ford is shown on the map, to the east of 'Humbles Staith'. The intention was to attack Newcastle on its vulnerable south side. The Royalist army was hurriedly forced to establish a defence at the ford and the two armies met in the battle of Newburn Ford on 28 August 1640. The Scottish army not only heavily outnumbered the English, but also occupied the higher and better ground north of the Tyne around Newburn itself. Although this much later map of 1763 does not actually mark the site of the battle, it illustrates the strategic position admirably. It also gives an impression of the contrast in topography between the two sides of the river. The Royalist forces were located south of the river on what was low-lying marshy ground, indicated by the featureless area called Ryton Haughs, whilst the Scottish army

not only commanded the high ground to the north but could also shelter behind field hedges. Cottages could be used as fire positions, and cannon were secreted in the woods to the east and north-east of Newburn. Despite the hasty construction of two 'sconces' or temporary fortifications to guard the ford, the English position was clearly at a distinct disadvantage. The Royalist general Lord Conway's choice of Stella Hall as his headquarters may have been influenced by the fact this was the most comfortable accommodation for miles around, but it placed his army in a strategically weak position. Having bombarded the Royalists from the higher ground across the river, including a gun platform established on Newburn Church, the Scottish forces were easily able to breach the defences at the ford and, after some brief skirmishes, the English cavalry and troops scattered in disarray. Their defeat led to the Scottish occupation of Newcastle. A major strategic outcome was the control of the town's coal trade and hence of London's coal supply by the invading Scots. Charles had little option other than to agree to pay the Scots daily expenses for billeting their troops while they occupied Newcastle. They eventually marched back to Scotland in 1641 once they had been paid some £300,000 for their 'brotherly assistance', according to the historian Eneas Mackenzie in his 1827 *History*.

In order to raise the necessary funds to meet the substantial costs that had been incurred, Charles was forced to summon the 'Long Parliament' that led to financial clashes but more fundamentally also to disputes over the respective rights of monarch and Parliament. In the subsequent Civil War the largely Puritan Parliamentarians found common cause with the Scottish Presbyterians in battling with the Royalists. Again, Newcastle found itself heavily involved. Unlike most other large towns, it sided with the Royalists, in part because the town and the wider area had numerous Catholic families but also because it sought to maintain the benefits of royal patronage and the protection of its coal monopoly. The key battle which helped to seal Newcastle's fate was in 1644 when the combined English Parliamentarians and Scottish Covenanters defeated the Royalists at Marston Moor (between York

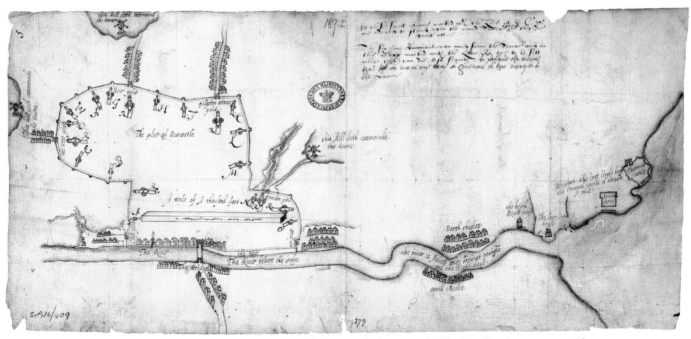

The 'plot' of Newcastle (1639) [NA] The 'owlde forte' is the Spanish battery which was repaired by Royalists in 1643 to provide casemates for guns.

and Wetherby). This effectively ended Charles's control of northern England and meant that Royalist troops could no longer come to the support of Newcastle.

Earlier in the year a large Scottish army, again led by Leslie – now Lord – Leven, had moved south across the border and left six regiments to besiege Newcastle. Sir Jacob Astley's rough sketch map shows the wider strategic position of Newcastle and the placement of cannon in relation to the walls and gates. Astley had been under no illusions about for how long Newcastle could hold out and pointed out the sites from which the town could be bombarded – 'this hill doth commande the town'. Despite the inability of the Royalists to send military assistance to the town's aid, Newcastle held out between February and October under the leadership of the mayor, Sir John Marley. Eventually, the Scottish army was able to breach a south-west section of its walls using artillery fire and mining, and the town was once again occupied by the Scots, reinforcing the 'ancient grudge.' The long and staunch defence that had been offered by the town earned it the somewhat inappropriate motto *Fortiter defendit triumphans* ('Triumphing by Brave Defence') from Charles. As with the earlier Scottish occupation, much damage and ill will resulted from the occupation. The Scottish occupiers also forced the surrender of Tynemouth and South Shields at the same time, with their complete control of the Tyne enabling them to use coal taxes to pay for their military incursions and occupation of Newcastle.

When Charles surrendered to the Scots in 1645, he was taken to Newcastle where it is thought that he was kept in Anderson House for some nine months while the Scots tried to persuade him to sign the Covenant. His refusal to do so led to his being taken to London and to the trial in which he was convicted of 'an unlimited and tyrannical power to rule according to his will, and to overthrow the rights and liberties of the people'. He was executed in Whitehall in January 1649.

One of the inevitable consequences of the Civil War for Newcastle was the disruption of its coal exports by blockades of the Tyne. The immediate beneficiary was Sunderland, which was staunchly Parliamentarian, like all the other large seaports except Newcastle and King's Lynn. Sunderland had long had its trade affected by Newcastle's monopolistic position but was now able to build up exports to London which otherwise would have been starved of coal supplies. While its coal exports were never on the scale of Newcastle's, this represented a significant dent in Newcastle's monopoly.

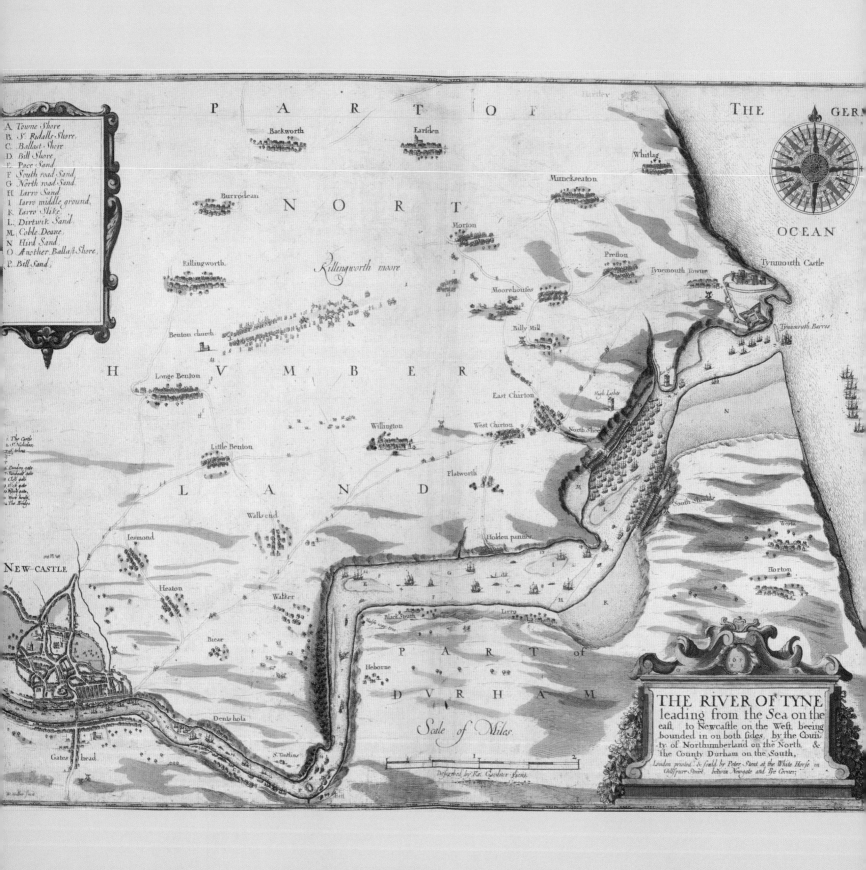

A. Towne Shore
B. St. Rdalls Shore
C. Ballast Shore
D. Bill Shore
E. Pace Sand
F. South road Sand
G. North road Sand
H. Iarro Sand
I. Iarro middle ground
K. Iarro Slike
L. Dirtwit Sand
M. Coble Deane
N. Hird Sand
O. Another Ballast Shore
P. Bill Sand

1. The Castle
2. St. Nicholas
3. ye towne
4. ...
5. Sandiga gate
6. Sandwich gate
7. Clife gate
8. Hich gate
9. ...
10. Kay house
11. The Bridge

PART OF THE GERM... OCEAN

P A R T   O F

N O R T H U M B E R L A N D

Backworth

Earsden

Whitlag

Munckseaton

Morton

Preston

Tynemouth Towne

Tynmouth Castle

Tynmouth Barres

Burrodean

Killingworth

Killingworth moore

Moorehouses

Billy Mill

High Lighte

Benton church

Longe Benton

East Chirton

West Chirton

North Sheele

Little Benton

Willington

Iesmond

Flatworth

Wallsend

Holden pannes

NEW-CASTLE

Heaton

Walker

Blacke Steath

Iarro

South Sheele

Horton

West...

Byear

Heborne

Dents hola

P A R T   of   D U R H A M

Scale of Miles

Gates head

S. Tantling

THE RIVER OF TYNE
leading from the Sea on the
east, to Newcastle on the West, beeing
bounded in on both sides, by the County of Northumberland on the North, &
The County Durham on the South,

London, printed, & sould by Peter Stent at the White Horse in
Gilsspurr Street, betwix Newgate and Pye Corner.

Described by Ra: Gardiner Gent:

1655

The erosion of Newcastle's monopoly

The most serious threat was to Newcastle's right to control of the entire navigable river and its trade.

A.W. Purdue, *Newcastle: The Biography*, 2011

The Hostmen of Newcastle had long exercised a variety of monopolistic controls to prevent trade competition from other towns. The case brought against Newcastle by Ralph Gardner was a late example of the animosity felt towards the town in its attempts to enforce its hegemony. Gardner had been imprisoned in 1653 for his refusal, against the wishes of the Newcastle Hostmen, to close his brewery in North Shields. In 1655 he petitioned Cromwell to open up trade on the Tyne, publishing a 200-page tract entitled *Englands grievance discovered, in relation to the coal-trade*. In it he castigated the town's magistrates and argued for freedom of trade and navigation on the Tyne.

Copies of the tract are extremely rare, and not all contain his map of the Tyne. The map was engraved by the distinguished engraver Wenceslaus Hollar and is very characteristic of his style: minute detail engraved with faint lines, reflecting Hollar's short-sightedness. It is an interesting map. It shows a rather distorted portrayal of the shape of the Tyne, with a dense cluster of masted ships at both North and South 'Sheelds' with a mere handful shown at Newcastle – intended, no doubt, to support Gardner's argument about the desirability of opening up trade on the Tyne. The clearest propaganda aspect, however, relates to the state of the river. Ships are shown aground on several sandbanks and two or three appear to be sunk in midstream. A large vessel is struggling to cross the bar at Tynemouth. The map also depicts the numerous small villages in the environs, some with spelling that appears to be guided by local pronunciation – Heborne, Jarro, Bicar, Horton, Westa.

OPPOSITE: Wenceslaus Hollar, *The River of Tyne* (1655) [BL] Hollar is credited with this map but the note 'Described by Ra. Gardner Gent' beneath the scale suggests Gardner had some input.

Hollar (1655) [BL] Detail. The map may have had a propaganda motive, but according to Howell, a century earlier it had been noted that 'verye nedfoll it is that there were at every pointe of sand or Rock a becone or a boye'.

The castle and fort at Tynemouth are shown dramatically, if inaccurately, and the High and Low Lights at North Shields are clearly depicted – one of their early appearances on maps. The outline of Newcastle clearly owes much to Speed's plan of half a century earlier, although Gateshead, with its long street hugging the riverbank and overlooked by St Mary's Church, must have been from Gardner's own observations. An interesting feature is the depiction of the racecourse on Killingworth Moor, which shows a long line of spectators cheering on horses and with the start and finishing poles clearly shown at each end of its length of close to three miles. Some 50 years later, in 1721, the racecourse was moved to Newcastle's Town Moor.

Despite the tract's title, it contains relatively little reference to the coal trade; instead, it has a mix of chapters covering history, shipwrecks in the river, depositions against the

Newcastle magistrates, and petitions to lower the price of coal and to establish a market in North Shields. Of particular interest in Gardner's tract, however, is the series of sketches reflecting his polemic against Newcastle's stranglehold on trade. One sketch shows telling images of the workings of Newcastle's hegemony, though not in the coal trade but concerning shipwrights: it shows a lone figure trudging off up-river saying 'To Newcastle for shipwrights'. A second shows an altercation between a 'shipwright of Newcastle' and a 'shipwright of England', presumably unwelcome as his tools have been scattered on the ground. A spectator shouts that he should be carried to Newcastle because he is a foreigner. The Newcastle shipwright shouts why should we not work in this river as well as you do in the Thames and elsewhere in England and Wales. A third sketch shows two shipwrights having to tramp off to London to seek work on the Thames where they will be 'kindly received and admitted to work'.

Of course, Newcastle's hegemony began far earlier than this. In the early thirteenth century King John granted the town the right to form trade guilds, which rapidly began to restrict various trades so that they could only be carried out in Newcastle. Later that century the town's mayor attacked North Shields to disrupt Tynemouth Priory's trade in coal. In 1530 Henry VIII restricted the export of coal from quaysides other than at Newcastle, effectively granting a trade monopoly to the town's burgesses. The Newcastle Hostmen had formed a cartel to exploit the monopoly and control the keels that transported coal to collier ships downriver. The dissolution of the monasteries had a major effect on the control of coal production and its export. Previously, most of the region's mines were owned by monasteries which tended to restrict production in order to keep prices high, but with dissolution the mines fell into the hands of private owners and production increased rapidly. The Hostmen gradually bought up leases in the coalfields, and in 1583 Queen Elizabeth let the mines in Gateshead and Whickham to two Newcastle merchants who then leased them to the Hostmen to create the 'Grand Lease', giving the Hostmen a virtual monopoly on the digging and export of coal, hence their becoming known as 'the Lords of

Coal'. In 1600, the Queen further entrenched their position by granting a royal charter to incorporate the Company of Hostmen. Significantly, in 1623, when a national statute forbade monopolies, Newcastle was specifically exempted. Royal privilege continued to favour the town as long as it played its strategic role in defending England against the Scots. With the Hostmen's power to block coal exports from any town and interrupt trade on the Tyne, and with the guilds' ability to prevent a range of industrial activity outside Newcastle, the town could exert a forceful control over its region. It was said in the seventeenth century that only ten men could sell coal throughout England. There were numerous Hostmen, but the Company was essentially run by a small inner elite of powerful rich men who both controlled coal and ran the governance of the town: a mere dozen or so filled the posts of alderman, mayor, sheriff and MP of Newcastle, and circulated these positions amongst themselves.

The Civil War did something to puncture the dominance of Newcastle since the blockade of the Tyne and the consequent shortage of coal in London prompted Parliament to encourage trade from the rivers Wear and Blyth, although this was never enough to fully replace the tonnage that had come from the Tyne. After 1660, Newcastle's trade in coal rapidly recovered, but it was no longer an effective monopoly as Sunderland continued to export coal, much to the wrath of the Hostmen who attempted unsuccessfully to impose a tax on the Wearside traders.

The Hostmen's cartel progressively weakened and was effectively broken by the end of the eighteenth century as improved transport enabled more and more coalfield areas to be developed in regions beyond the North East. The fortunate proximity of North East coal deposits to the Tyne therefore lost its unique significance. Canal building, and particularly the growth of railways in the nineteenth century, undermined the dominance that the Tyne had previously exerted. There is an irony that the important role which the North East played in the innovative development of steam engines and a national railway system had a major negative impact on the region's domination of the coal trade.

Shipwright's map [SANT] Howell's description of this as a 'cruder map' seems somewhat misplaced.

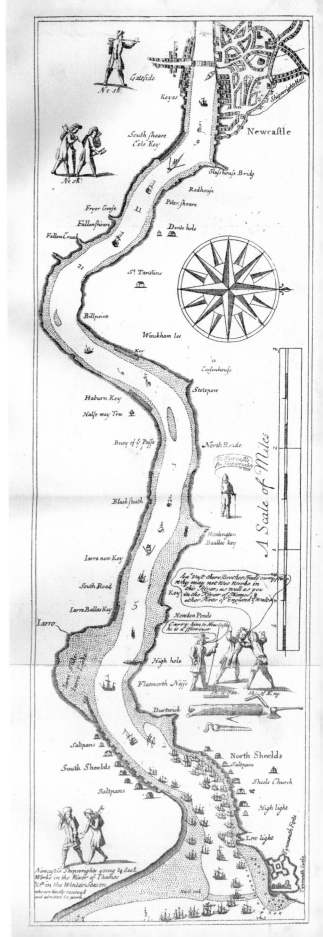

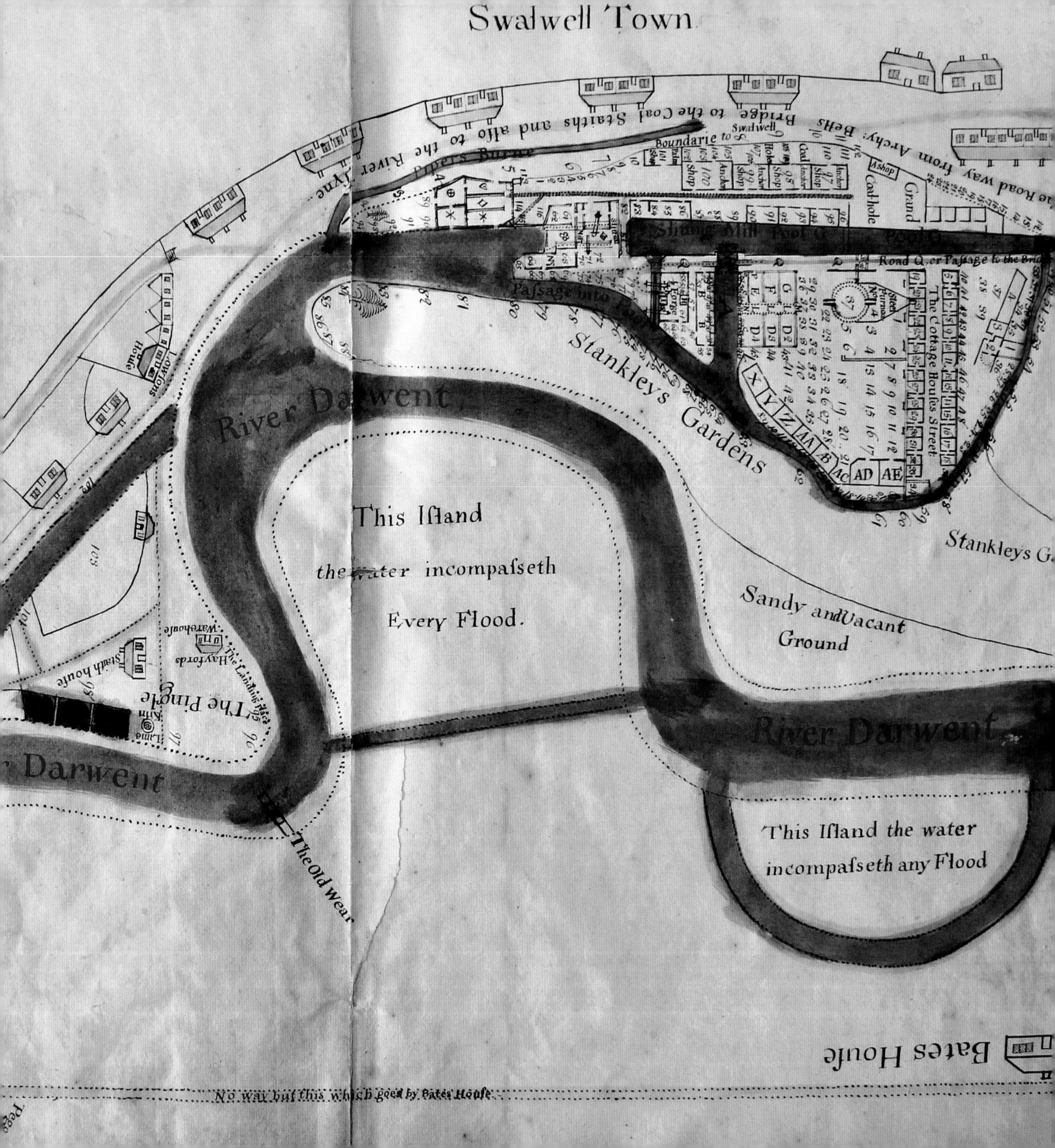

Swalwell Town

River Tyne

Bridge to the Coal Staiths and also to the River Tyne

Boundarie to

Bridge to the Coal Staiths and also to the River Tyne

the Road way from Archy Betts

Swalwell

Slitting Mill Rost

Road Q. or Passage to the Bridge

Passage into

Stankleys Gardens

River Darwent

Lowfons

This Island

the water incompasseth

Every Flood.

Stankleys Gardens

Sandy and Vacant
Ground

Stankleys Ga

Staith house

The Pingle

The Hayfords

The Landing Pla

Warehouse

River Darwent

River Darwent

The Old wear

This Island the water
incompasseth any Flood

Bates House

No way but this which goes by Bates House

1718

Crowley's early iron works

It is hard to believe that there could have been any larger industrial unit in the country at this time.

M.W. Flinn, *Men of Iron*, 2018

Arthur Young goes even further than Flinn, writing in 1770 that Crowley's is 'supposed to be the greatest manufactory of its kind in Europe'. He may well have been correct for, two decades earlier in 1754, the Swedish traveller R.R. Angerstein (sometimes accused of being an industrial spy) visited the Derwent Valley and provided a detailed account of Crowley's iron works at Winlaton, Winlaton Mill and Swalwell, the latter shown in the hand-drawn but detailed sketch map of 1718. What he failed to record, however, was that this enterprise was already well over half a century old and was probably the largest in Europe. Angerstein may have been somewhat jaundiced by the apparent lukewarm reception that he was

given, complaining that the clerks at Swalwell 'have nothing else to do than to make sure that no stranger enters the works . . . and starts talking to the workers'. He was clearly impressed, however, by the scale of the operation, especially at Swalwell as evident on the map, as well as its labour relations and its annual profit of £11,000. Yet today there is barely any trace of what was, arguably, a central constituent of Tyneside's early start in coal-based industrial activity.

In the 1680s, the London-based nailmaker, Ambrose Crowley III, had established an ironworks at Sunderland, breaking the virtual monopoly of the West Midlands. Sunderland offered considerable advantages for iron-making: it could import iron ore from a variety of locations, there was a plentiful supply of coal, labour could be found and accommodated and, as he delighted in observing, he could transport the finished product by sea to his shop in London more quickly than the

OPPOSITE: *A draught of Esq. Crowley's Works beginning at the High Dam above Winlaton Mill and ending at the River Tyne* (1718) [TWA] Note the unusual orientation. The left-hand side is north and 'Swalwell Town' printed at the top of the map is to the east.

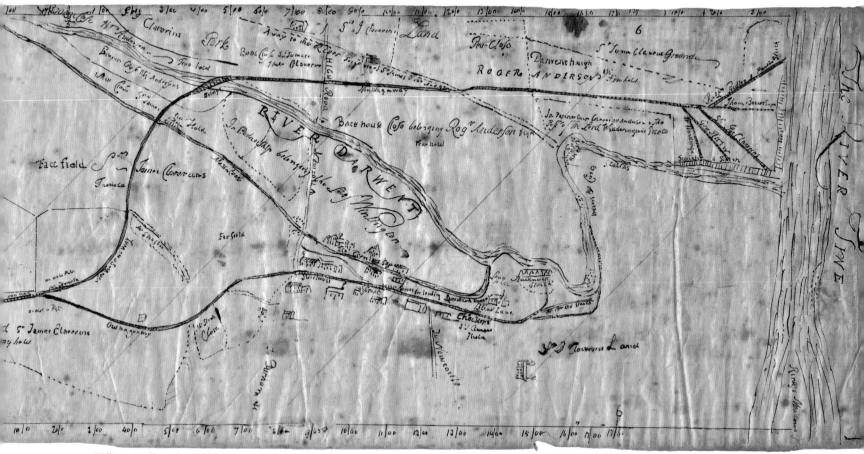

William Laidler, *The Derwent Ways* (1722) [BM] The many staithes in the lower Derwent and at Swalwell confirm this area's status as one of the principal river ports of the Tyne in the early eighteenth century.

same product could be transported there from Birmingham. But, having successfully established his Sunderland factory, around 1682 Crowley sought an alternative location, partly because of local hostility to the foreign element in his workforce. In 1691 he leased a disused corn mill at Winlaton, situated about a mile south of the Tyne, turned this into an ironworks and succeeded in securing contracts from the Royal Navy. Later in the decade he developed a completely new works at what became Winlaton Mill, 300 feet below Winlaton village, next to the fast-flowing Derwent river. Apart from access to local coal, other advantages here were the availability of charcoal from Chopwell Woods, easy access to staithes at

Blaydon and the existence of an earlier waggonway, constructed to carry coal from Durham mines to the Tyne.

Crowley also built a warehouse at Blaydon where bar iron was landed and the finished goods exported. However, in 1702 local competition emerged with a new partnership which was led by a Newcastle merchant and former Sheriff, Edward Harrison, and created to establish a new ironworks at Swalwell on the east bank of the Derwent close to its confluence with the Tyne. It is likely that Harrison had worked as an agent for Crowley. This new partnership leased three disused corn mills and started to attract a workforce, some of it from Crowley's works. Crowley's response was swift: by 1707 he had bought out this short-lived new enterprise to create a much larger industrial complex.

Local transport links were vital to Crowley's success. The 'old' waggonway shown on the map of 1722 passed through Swalwell to long-established staithes on the Derwent just to the east of the town. A note on the 1718 map says, 'These black places are coal staiths'! Due to the local rivalry of the Clavering and Cotesworth families, a new waggonway branch was built by the former to pass to the west and cross the Derwent before heading to a group of staithes on the Tyne just to the west of their confluence. But Crowley appears to have been unmoved by these rivalries. From 1707 until his death in 1713, Crowley made substantial investments at this location, and these were continued by his son. The administrative functions of the firm were moved to Swalwell and, by the first quarter of the eighteenth century, the three sites contained an immense range of forges, furnaces and workshops. Lighter articles such as tools and nails were made at Winlaton, while heavier forgings such as anchors, chains and pumps were made at Swalwell. The structure of the operation here is immediately apparent from the detailed map and consisted of a series of neighbouring but relatively small workshops, smithies, etc. The reference numbers relate to the different specialised functions of individual buildings. Angerstein listed over 60, including no less than 22 for the forging of hoes with three workmen in each, ten workshops for the manufacture of small nails and two for ship's nails, three anchor forges with six workmen in each (in fact, the map shows four, numbered from 97 to 100), two steel furnaces (one is numbered 37 on the map), a grinding mill, a slitting mill and a plating forge. Metal objects of all kinds were made, with the navy being a major customer, but the initial specialisation on nails continued, with 108 different kinds being manufactured. It has been estimated that almost 1,000 workers were employed, some recruited locally but the majority initially coming from further afield, especially Yorkshire and the West Midlands.

Proximity to Newcastle was also crucial to Crowley's expanding business through its control of customs duties, trans-shipment and wharfage dues, and the firm maintained admin-istrative functions in the town. The emergence of an integrated regional economy with Newcastle at the centre is also shown in its provision of banking services for the firm, including the supply of cash for the payment of weekly wages.

The firm's operations expanded further in 1735 with the acquisition of works in the Team Valley where another steel furnace was built in 1740. The large size of the business continued for most of the eighteenth century, fuelled by the growth of the navy and the demands of frequent warfare. By the end of the eighteenth century the firm became Crowley, Millington & Co., and in the early 1830s it was described by Thomas Oliver as 'one of the largest establishments in the kingdom'. However, the peace of 1815 led to a significant reduction in demand and the old Winlaton works ceased operations in 1816. The firm's dominance was also challenged by the growth of competition, especially from the Hawks family whose naming of their factories (New Woolwich and New Deptford) on the Tyne at Gateshead signalled their ambition. This firm also expanded into a wider area of production, for example securing the contract to build the renowned High Level Bridge between Newcastle and Gateshead.

The Crowley enterprise was also significant as being probably the first large industrial enterprise to concern itself with the social welfare of its workers. This was not only in the provision of housing. The Crowley works had its own 'laws' and social security system, including a system of sick pay, a health service, a school and widow's pension, all this more than 200 years before such things became available nationally. Crowley referred to 'my people' and, as a large-scale exercise in industrial paternalism, the complex was probably unique in Europe. The system was funded by both employer's and employees' contributions, but continuing employment was subject to adherence to rigorous labour discipline, partially ensured through a system of informants. Furthermore, the strong dependence on navy contracts inevitably meant a fluctu-ating demand for labour, and while key workers were paid retaining wages, day labourers were afforded no such luxury. However, the extinction of the Crowley family line in 1782 led to its charitable institutions being abandoned.

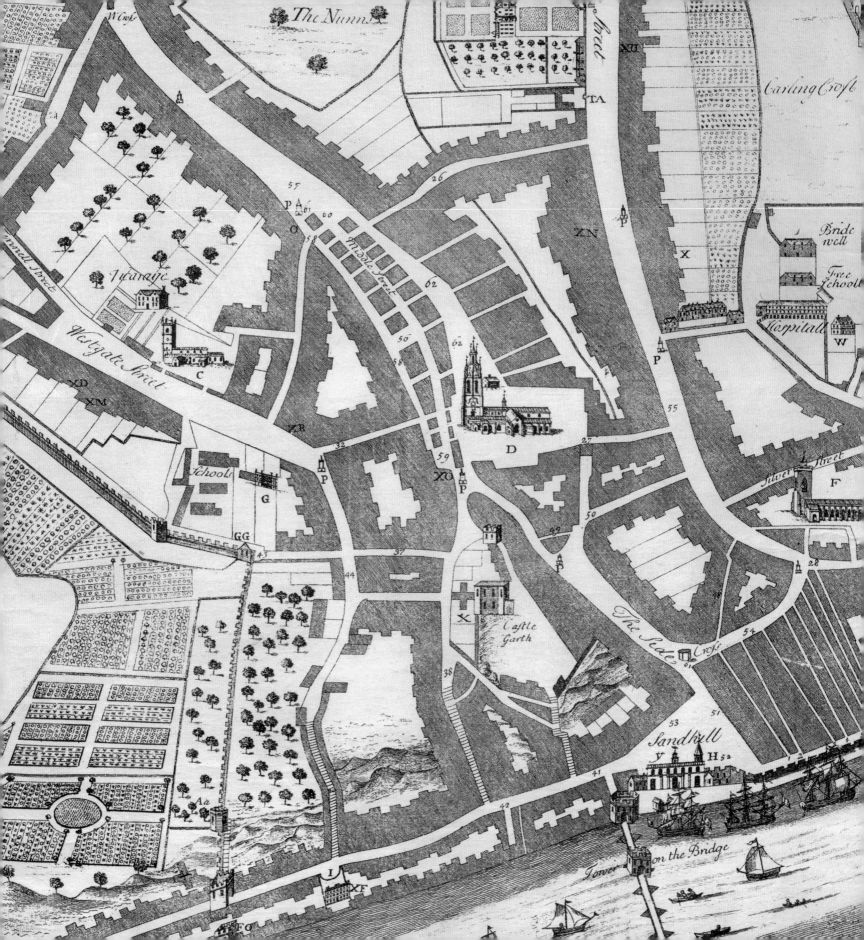

1723

Newcastle's first surveyed plan

Mr James Corbridge having . . . with great labour and pains drawn a Plan of this Town describing the several streets, wards & other remarkable places therein. It is ordered that five guineas be now given him towards engraving the same.

Newcastle Common Council, 27 March 1724

As with many early cartographers, only sketchy details are known about James Corbridge. When and where he was born is unknown, although his name suggests a northern origin. He was living in Newcastle's Pilgrim Street in 1723 when he surveyed the town. His plan is important, as it was the first to be produced from an actual survey since Speed's in 1610 (published 1611) and is full of detail. Its most immediate impact is the set of 26 views that surround its margins and illustrate the town's public buildings and major private houses.

This design was probably inspired by James Millerd's 1673 plan of Bristol which is thought to be the first English plan to incorporate detailed views of buildings around its margins. Other cartographers used the design on later eighteenth-century town plans, for example the large-scale plans of Leeds and York by John Cossins and the sequence of plans of Manchester by Casson and Berry.

Corbridge's illustrations include views of the four Newcastle churches and of St Mary's in Gateshead (although that is not depicted on the plan). Other public buildings include the Exchange in Sandhill and trade halls such as the Barber Surgeons and the Ropers. His reference system is somewhat chaotic: as well as the alphabetic references to the views around the margins, there are extensive lists of references along the bottom of the map that include the gates and towers of the defensive walls, the lists of lanes and markets and the lists of

OPPOSITE: James Corbridge, detail from *Newcastle upon Tyne* (1723) [TWA] A classic medieval core with church and elongated market place where former temporary stalls had been converted to permanent buildings by the early eighteenth century.

trades and halls. References to the lanes and markets are numerical and easy to identify on the map. However, the alphabetic references to the gates and towers and to the companies and halls are confusing, some duplicating those on the illustrations: for example, 'A' signifies both the Masons' Hall and William Blackett's house; 'T' shows both the Millers' Hall and Trinity Hall; and 'X' identifies not only the Shipwrights' Hall and the house of William Soursby, but also a large gate or tower at the right-angular section of the defensive wall at Pandon, which may signify the Sally Port or the Wall Knoll Tower. Corbridge mistakenly names 'chares' as 'chairs' which suggests that he may not have been local and that he simply translated what he was told into a word with which he was familiar.

Nevertheless, despite such glitches, this is an invaluable map, not least because it was surveyed at a time when the town was at the start of major changes. In the seventeenth century, many of the rich merchants and prominent citizens had spacious houses close to the river in the Side and the Close. However, from the early eighteenth century onwards, many of the richer citizens moved onto the plateau to live in Westgate Street and Pilgrim Street. Corbridge's map shows the location of some of these grand houses. Examples include: Sir William Blackett's house (later called Anderson Place) at the head of Pilgrim Street; those of Fenwick Lambert, Alderman Fenwick, Thomas Brumell and William Soursby situated lower down Pilgrim Street; and those of Lady Clavering and Richard Swinburn on Westgate Street. But the Mayor's House remained on the Close.

Some of the other references in Corbridge's map are also of interest, for instance the dissolved Blackfriars is shown as housing the companies of skinners, tailors, bakers, brewers, tanners, cordwainers, saddlers, butchers, smiths, fullers and dyers. Indeed, Corbridge's full list of trades gives a good indication of the economic activities that then dominated the town. Critically, given the significance of coal exports and the river, the Hostmen's Hall is shown accommodated within the Guildhall on Sandhill, and Trinity House is shown in the eastern section of the chares. Of further interest is Corbridge's identifi-

cation of 'pants' (the North East's term for water pumps or conduits): three are shown in Pilgrim Street, two in the Close, two in Newgate Street and one in Westgate Street. Other small symbols probably also indicate pants, although they are not marked as 'P'. An entertaining feature of the map is its playful views to the east of the town. Small figures are shown flying a kite in St Ann's grounds, cattle and sheep are grazing in the fields outside the walls, horsemen are galloping in a field just north of the long ropery, and a horse-drawn coach with its rider flourishing his whip is hurrying along the road towards Shields.

The version of Corbridge's map shown here comes from a lithographed copy published for the Newcastle Society of Antiquaries by Angerer & Goschl in Vienna in the 1880s. Although subsequently frequently reproduced, the original map is very scarce. In researching Corbridge's career, the cartographic historian Raymond Frostick, for example, could find only six copies. This may be because most large town plans suffered inevitable damage, but it is equally likely that Corbridge was not very successful in attracting subscriptions. He advertised the plan on a number of occasions in the *Newcastle Courant*: in January 1723 he claimed, incorrectly, that 'the plates for printing the Plan of Newcastle upon Tyne, lately survey'd and Taken by James Corbridge, are now Engrav'd. and some of them already Printed, and the rest will be finish'd in a short Time'; in December the same year he advertised the plan and looked for subscriptions; and in October 1724 he advertised that 'the Plates . . . are now Engrav'd, and some of them already Printed, and the rest will be finish'd in a short time'. By that time, however, he had left Newcastle and gone to Norfolk, a move that may have been prompted by poor sales in Newcastle and the better prospects that he saw in Norfolk. Indeed, he produced a very fine large-scale prospect of Great Yarmouth in 1725, followed in 1727 by a two-sheet plan of Norwich, very similar in layout to the

OPPOSITE: Corbridge (1723) [TWA] Even in 1723 the amount of open space within the walls is particularly notable.

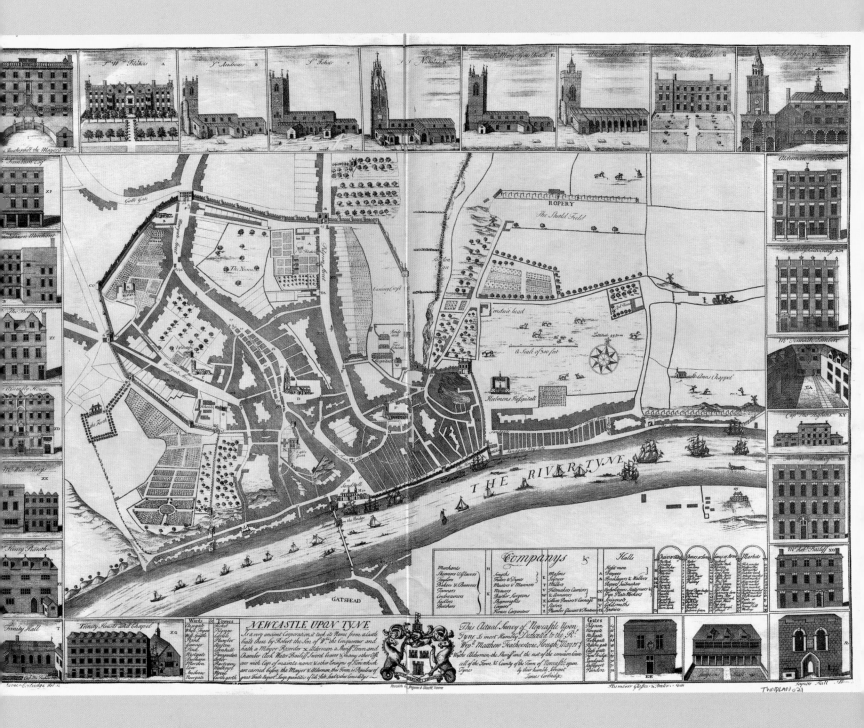

NEWCASTLE UPON TYNE

THE RIVER TYNE

GATSHEAD

Companys

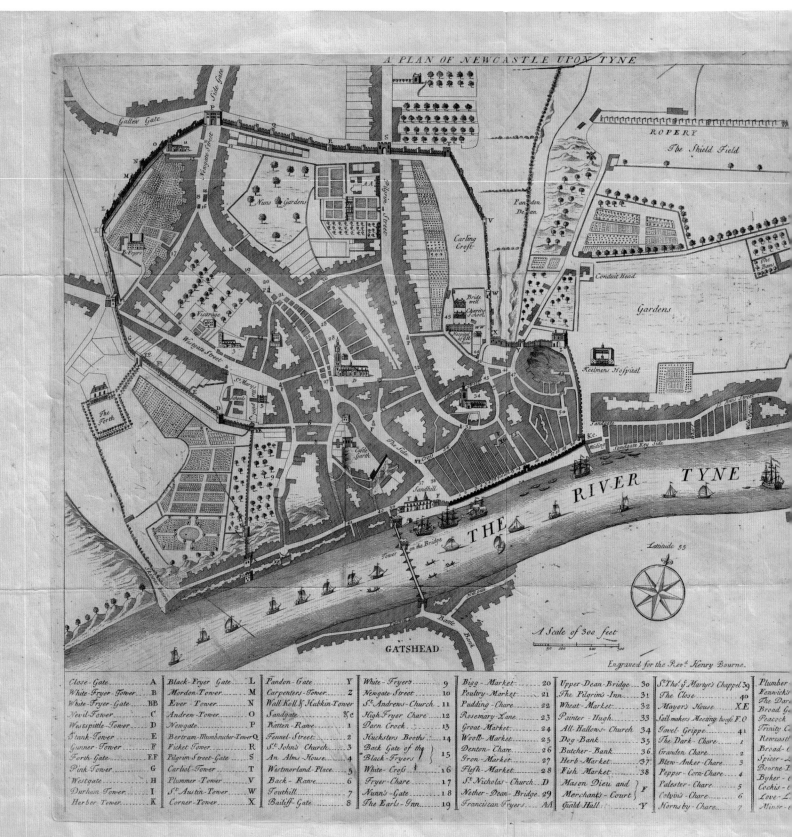

A PLAN OF NEWCASTLE UPON TYNE

ROPERY

The Shield Field

Gardens

Keelmens Hospital

THE RIVER TYNE

Latitude 55

GATSHEAD

A Scale of 300 feet

Engraved for the Revd. Henry Bourne.

Close-Gate	A	Black-Fryer Gate	L	Pandon-Gate	Y
White-Fryer-Tower	B	Mordon-Tower	M	Carpenters-Tower	Z
White-Fryer-Gate	BB	Ever-Tower	N	Wall Koll & Habkin-Tower	
Nevil-Tower	C	Andrew-Tower	O	Sandgate	&c
Westspittle-Tower	D	Newgate	P	Ratten-Rawe	1
Stank-Tower	E	Bertram-Mumboucher-Tower	Q	Fennel-Street	2
Gunner-Tower	F	Ficket Tower	R	St. John's Church	3
Forth-Gate	EF	Pilgrim-Street-Gate	S	An Alms-House	4
Pink-Tower		Carliol-Tower	T	Westmorland-Place	5
Westgate	H	Plummer-Tower	V	Back-Rawe	6
Durham-Tower	I	St. Austin-Tower	W	Touthill	7
Herber-Tower	K	Corner-Tower	X	Bailiff-Gate	8

White-Fryers	9	Bigg-Market	20	Upper-Dean-Bridge	30
Newgate-Street	10	Poultry-Market	21	The Pilgrim's-Inn	31
St. Andrews-Church	11	Pudding-Chare	22	Wheat-Market	32
High-Fryer Chare	12	Rosemary-Lane	23	Painter-Hugh	33
Darn Crook	13	Groat-Market	24	All-Hallows-Church	34
Hucksters Booths	14	Wooll-Market	25	Dog-Bank	35
Back Gate of the Black-Fryers	15	Denton-Chare	26	Butcher-Bank	36
White-Cross	16	Iron-Market	27	Herb-Market	37
Fryer-Chare	17	Flesh-Market	28	Fish-Market	38
Nunn's-Gate	18	St. Nicholas-Church	D	Mason Dieu and Merchant's-Court	F
The Earls-Inn	19	Nether-Dean-Bridge	29	Guild-Hall	Y
		Franciscan Fryers	AA		

St. Thos. ye Martyrs Chappel	39	Plumber
The Close	40	Fennicks
Mayor's House	XE	The Dark
Sail-makers Meeting house	F.O	Broad Ga
Gavel-Grippe		Peacock
	41	Trinity C
The Dark-Chare	1	Newcastl
Granden Chare	2	Broad-L
Blew-Anker-Chare	3	Spicer-L
Pepper-Corn-Chare	4	Bourne
Palester-Chare	5	Byker-
Colvin's-Chare	6	Cockis-
Horns-by-Chare	7	Love-L
		Minsr-

D/NCP/2/6

LEFT: Henry Bourne, *Newcastle upon Tyne* (1736) [TWA] Bourne did not live to see the publication of his *History of Newcastle*, dying at the early age of 37.

RIGHT: Detail of market area from Henry Bourne, *Newcastle upon Tyne* (1736). The large building on the left is the Presbyterian chapel connected to the Church of Scotland, built around 1715.

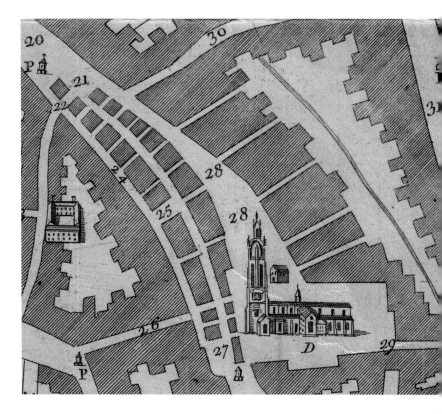

design that he used for Newcastle. He also published the first large-scale map of Norfolk.

Corbridge's plan of Newcastle had a further life: it was used (though without attribution to Corbridge) in Henry Bourne's *History of Newcastle* which was published posthumously by his children in 1736. The plan was shown at about half the size of the original and did not include the marginal illustrations. Also, it does have some minor alterations from the original: it has a different configuration of vessels on the Tyne and the flags of some appear not to be blowing in the prevailing direction; it excludes the various figures and scenes in the eastern outskirts; and it correctly uses the term 'chare' for the narrow lanes cascading down from Dog Bank to the Quayside and elsewhere in the town. Its reference system is still somewhat confused, but sensibly it makes no attempt to show private houses by complicated and duplicated keys. Essentially, however, Bourne's map is identical to Corbridge's.

1745

Town panoramas

This truly glorious panorama shows in exquisite depth and detail one of England's greatest northern towns in all its full mercantile splendour.

Arader Galleries, Philadelphia, 2020

Panoramas and bird's-eye views were a favourite way of depicting towns until the advent of the formally surveyed plans that became common in Britain in the later eighteenth century. Panoramas have the advantage of showing a town as it would be seen by an observer on the ground, but also the disadvantage that, while they can portray the immediate foreground, they can show more distant elements only as crowded perspectives and so cannot delineate the geometry of a town and the lines of its streets. Bird's-eye views can do this somewhat better if their viewpoint is from a sufficiently high angle, but that robs the view of any realistic portrayal of topography if the town sits on anything other than a flat plane. Nevertheless, such views and panoramas can provide helpful and attractive prospects. The temporal juxtaposition of Corbridge's map of Newcastle (*see* 1723) and Samuel Buck's panorama (first edition *c*.1723) demonstrates the two forms of depiction. Intriguingly, both Buck and Corbridge were awarded five guineas at the same meeting of the Newcastle Common Council in March 1724, clearly showing significant interest in depicting the town but suggesting that there was also some uncertainty in how Newcastle could be best represented.

The best-known panoramas were produced by Samuel and Nathaniel Buck who drew no fewer than 87 views of towns. Samuel, the older of the two brothers, began by drawing his 'prospects' on his own, encouraged by the Leeds-based antiquarian Ralph Thoresby and later by members of the Society of Antiquaries of London, notably William Stukeley.

DETAIL OPPOSITE AND OVERLEAF: Samuel and Nathaniel Buck, detail from *The South East Prospect of Newcastle upon Tyne* (1745) [GAC] Amongst its many attractions is possibly the best representation of the medieval bridge.

Samuel's initial drawings comprised ten huge perspectives called 'the First Series'. They focused almost entirely on towns in the North, no doubt reflecting Buck's origins, given that his hometown was probably Richmond in Yorkshire. Newcastle was one of the towns he included. The first edition of this prospect measures 1195 mm by 495 mm and was published around 1723, along with Durham, Stockton and Sunderland.

In 1724 and 1725 Stukeley took Samuel on his travels around the country to assist with drawings that appeared in Stukeley's *Itinerarium Curiosum*. Thereafter, from 1728, Samuel worked with his brother, Nathaniel, to produce their extensive array of prospects, views and depictions of churches, castles and antiquities. These became known as the 'the Principal Series' and they include this prospect of Newcastle, published in 1745.

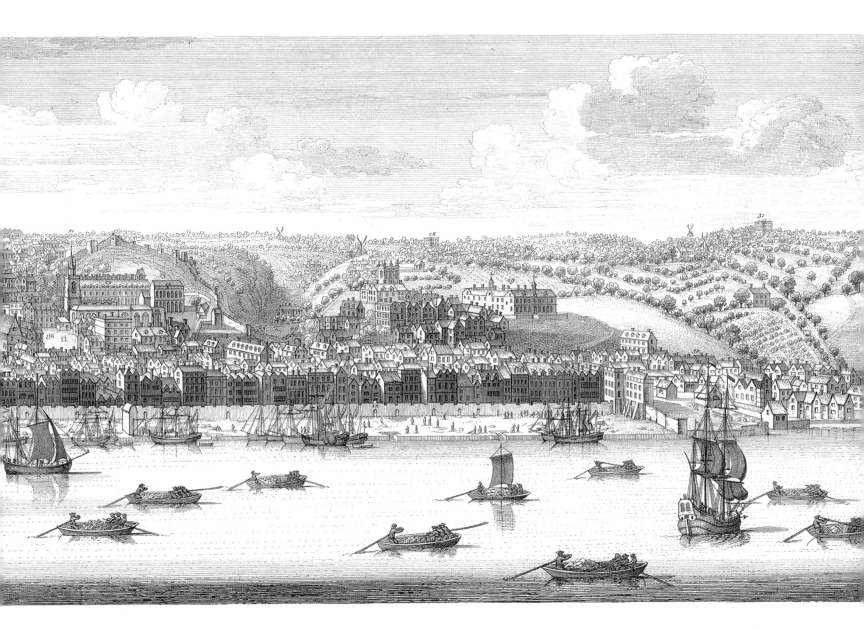

These prospects were smaller than those of the First Series, although still large, typically 305 mm by 775 mm. Newcastle's panorama was drawn from a vantage point on the tower of St Mary's in Gateshead looking directly across the river to the plateau on which the upper town had developed. It is this which gives a somewhat false impression of the town's topography since the steepness of the banks above the Tyne is not

really captured. However, Pandon Dene is shown as a deep ravine running down to the river. The panorama offers a striking impression of the town at that time. Above the bridge the mayoral barge is depicted, while below the bridge there are numerous keels taking coal downstream, as well as sea-going ships on the river and moored at the Quayside. The quay is crowded with men moving and loading cargoes. Behind the

John Brand, *View of Newcastle upon Tyne taken from the south side of the river* (1783) [AUTH] The coal transport
activity in the foreground contrasts with the prime river frontage of merchants' houses in the Close.

long stretch of the quay, the town walls are shown in their
repaired condition ready for the expected Jacobite Rebellion
of that year. The medieval bridge is carefully depicted with its
houses, gates and chapel. The castle and St Nicholas Church
are accurately drawn. In the town's environs, windmills are
shown on the skyline, and in the distance villages that were to
become suburbs of the town are identified: 'Jasment', Heaton
Hall and Elswick. The textual account of the town calls it 'the
great Emporium of the Northern parts of England' and empha-
sises its trade in coals, lead, salt, grindstone and glass, fully
justifying the busy activity on the Quayside.

There are numerous other later panoramas of the town,
none of which effectively capture the gorge-like nature of the
Tyne. One of the better ones is a view that was used by John
Brand in his *History*. It depicts Newcastle in 1783, seen from
a viewpoint above Gateshead to the west of the bridge, a
slightly different angle from the Bucks' vantage point. The
foreground has a lively view with a collier navigating a coal

truck on a waggonway and with keels on the river. All Saints,
the castle and St Nicholas are prominent on the skyline and,
facing the river, the Mansion House is dominant amongst the
long line of substantial houses on the Close. In the west, clouds
of black smoke are belching out of one of the major glass works
on the river front. The medieval bridge had been swept away
in the 1771 flood and the view shows its successor with its
stone arches, across which a horse is being ridden.

Despite increasingly sophisticated mapping techniques,
panoramas remained a popular way of depicting the city until
it was replaced by photography well into the nineteenth
century. Unlike the accurately surveyed map, the panorama
had the advantage of providing the viewer with the illusion
that she or he was physically present at the scene being
depicted. The *Illustrated London News* developed a distinctive
tradition of graphic illustrations for key events. Newcastle's
townscape at the time of Queen Victoria's Golden Jubilee
offered a suitable subject and is represented in typically

The Illustrated London News, Newcastle upon Tyne, 1887 (1887) [AUTH] Drawn by the prolific
Thomas Sulman who pioneered the use of hot-air balloons to draw bird's-eye views of cities.

dramatic fashion. The river and Quayside activity dominate the right-hand foreground, with the gas works and smoke-stacks of Abbot's engineering Park Works in Gateshead being proudly displayed. Modernity and invention are celebrated in the prominence given to the juxtaposition of the High Level Bridge, with two locomotives, and the Swing Bridge. The changing nature of at least one aspect of river trade is shown by the five keels being towed upstream by a small steam tug. New commercial buildings have arisen to the east of Sandhill and the Guild Hall to replace those destroyed by the Great Fire of 1854. The Castle Keep and John Stokoe's 1812 Moot Hall courthouse stands boldly dominant, but it is also possible to pick out how the railway lines have tightly enclosed the Castle Garth to west and north. However, one of the most curious features of the illustration is the apparently vast cleared area between All Saints Church and the distinctive Sallyport building. This is the area of Stockbridge which was indeed scheduled for slum clearance by the Medical Officer of Health, Henry Armstrong, in maps of 1876, but the 25-inch OS map of 1894 shows that only a small area had in fact been cleared by that date. This reminds us that, attractive and spectacular as an artist's panorama may be, there is no substitute for the accurately surveyed map if a precise depiction of reality is what is desired.

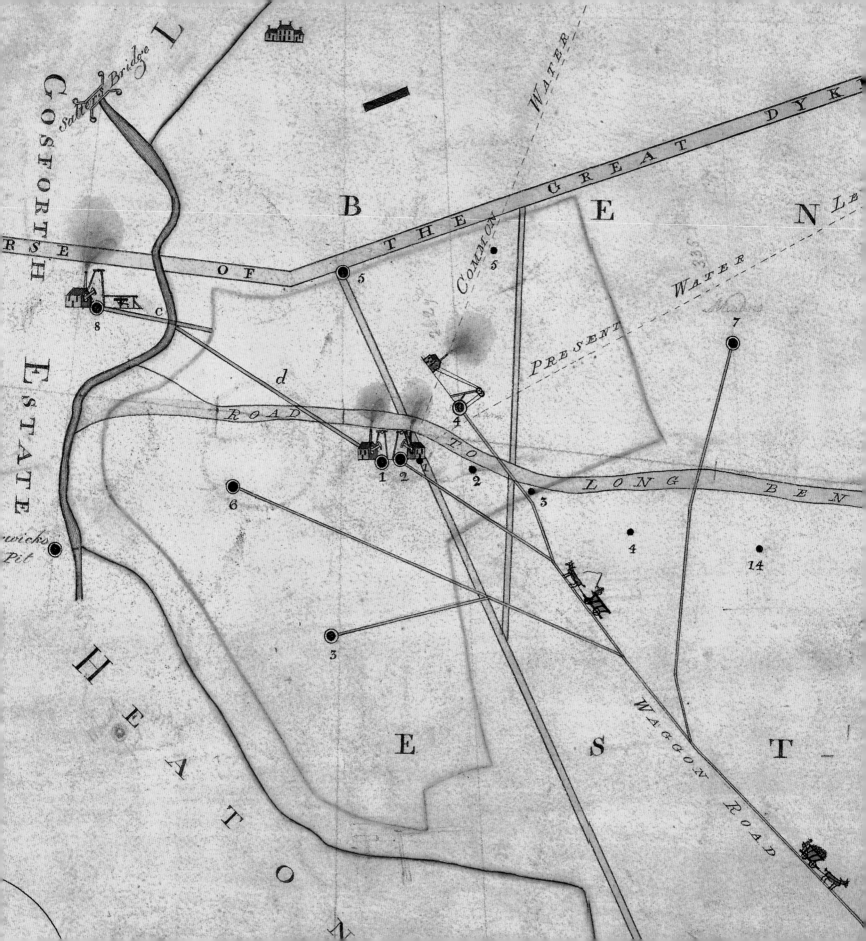

1749

Waggonways and early railways

These waggon-ways . . . may vie with some of the great works of the Roman Empire.

H. Bourne, *The History of Newcastle upon Tyne*, 1736

The map of the Earl of Carlisle's Longbenton estate in 1749 shows the waggonways associated with coal exploitation in fine detail. Eight pits, numbered and indicated by a black dot with enclosing circle, were individually serviced by branches which then combined for the route to staithes at Walker. The detailed drawings show that haulage was by individual horses, controlled by one man and with a brake lever to control the rate of descent. Horses were swapped around for the run down to the staithes or taking empties back to the pits. Longbenton colliery appears to have been profitable as a viewer's report of 1850 tells us that at Meadow Pit alone (number 7 on the map)

there were 37 horses who had conveyed 123 waggon loads of coal in 1750–51. Overall, the exploitation of the estate for coal was advanced for the time, with three steam-driven pumping engines installed there. The engine at number 4 on the map appears rather different and may well have been a stationary haulage engine operating on a short stretch of track at this stage. The map is suggestive of some of the innovations taking place in coal mining and transport even in the pre-locomotive era.

Up to the mid seventeenth century there was little significant coal working more than three miles from the Tyne or its navigable tributaries. The cost of carrying coal to the river in 'wains' (carts usually pulled by a combination of oxen and horses) was extremely high, and routes became impassable in wet weather. A partial solution to the cost issue came with the development of wooden waggonways, enabling larger quantities of coal to be transported more quickly and relatively free

OPPOSITE: Detail of *Plan of Long Benton Estate* (1749) [MIN] The operation of the waggons is deftly illustrated by the use of a brake lever on the descending waggon whilst the horse pulling uphill is encouraged by a whip.

from the effects of bad weather. The impact can be gauged by figures from the Tanfield Moor colliery where transport by a horse-drawn single waggon yielded a profit of £1 17s 6d per ton of coal, but transport by waggonway increased the profit per ton to £5 5s 9d.

Before around 1712, waggonways were usually just single track with passing places, but subsequently double-track lines developed. They also began to require significant land preparation, hence there was a growth of specialist land surveyors and engineers. However, waggonway development was subject to obtaining 'wayleaves' (the right to cross other people's property), and this of course became increasingly expensive (and profitable!): Lady Clavering, for example, obtained 2s 6d per ton of coal for allowing a waggonway to cross just 300 yards of her property on Tanfield Moor.

The first recorded waggonway in the North East is noted in William Gray's *Chorographia* (1649), referring to the introduction of a waggonway linking coal working at Cowpen to the River Blyth – 'waggons with one horse to carry down coals, from the pits to the staithes'. A second early example was at Whickham which had functioning waggonways in the 1620s. Other early ways to the Tyne were constructed from Fawdon colliery in 1656, from Crawcrook in 1663 and from Stella, also in the 1660s. In 1669 the Ravensworth Way was built from Ravensworth and Blackburn collieries to staithes on the River Team, and another in the 1670s from Winlaton to the Tyne. These early waggonways were very short and served collieries only a mile or two from the river. But they were locally important, for example at Cullercoats where 36 salt pans used 15,000 tons of coal to produce 2,200 tons of salt in 1708. Overall, from the mid seventeenth century through to the early nineteenth, Leslie Turnbull's comprehensive study identifies 36 main waggonways with 63 branches in Northumberland, along with 31 main waggonways with 36 branches or extensions running up to the south bank of the Tyne.

The extract from J.T.W. Bell's map of the Newcastle Coal District east of Newcastle shows the number and complexity of the waggonways that had developed with the expansion of the coalfield in the lower Tyne area. Technical innovations in

dealing with water ingress allowed deeper mining, but innovations in the waggonways were equally significant. In the later eighteenth century, cast-iron strips started to be laid on top of wooden rails to prevent excessive wear, and towards the end of the century metal rails started to predominate. More efficient braking systems were developed, as too were inclined planes where the weight of the loaded descending waggons could be used to haul up a similar number of empties. The use of stationary engines for haulage purposes also increased. Together with the metal rails, this allowed much heavier loads to be carried and led to the coupling of multiple waggons to form trains, as shown in the detail of Bell's map at Percy Main. This line from Seghill Colliery down to the staithes at Hay Hole (later the site of Northumberland Dock as printed on the map) was worked by two stationary engines, with the one at Percy Main hauling up empty waggons from the staithes.

One of the most significant lines in the later phase of waggonway expansion was the Wylam waggonway, following the opening of pits by the Blackett family in 1748. Coal was taken five miles to load keels at Lemington (*see* 1802), with a branch to Newburn also constructed. The particular significance of this line is that it was here where the region's first experiments with steam locomotives took place in 1814–15 when William Hedley built two locomotives. An equally important waggonway, however, was from Killingworth Moor to Willington Quay on the Tyne, built in 1765. George Stephenson's first locomotive, the *Blücher*, operated on this line and his 4 foot 8½ inch gauge became the standard for the entire country.

The origins of steam locomotives are much earlier than is usually recognised and did not take place on Tyneside. The Cornish engineer Richard Trevithick had built the first mobile steam engine in 1802 for Coalbrookdale, while a second was built in 1804 for the Pen-y-Darren ironworks in Merthyr Tydfil. Although this locomotive was so heavy that it broke the wooden rails, it attracted the attention of Christopher Blackett of Wylam. The successful adoption of steam locomotives was

OPPOSITE: *Plan of Longbenton Estate* (1749) [MIN] Pits are denoted by an encircled dot, other numbered dots are boreholes.

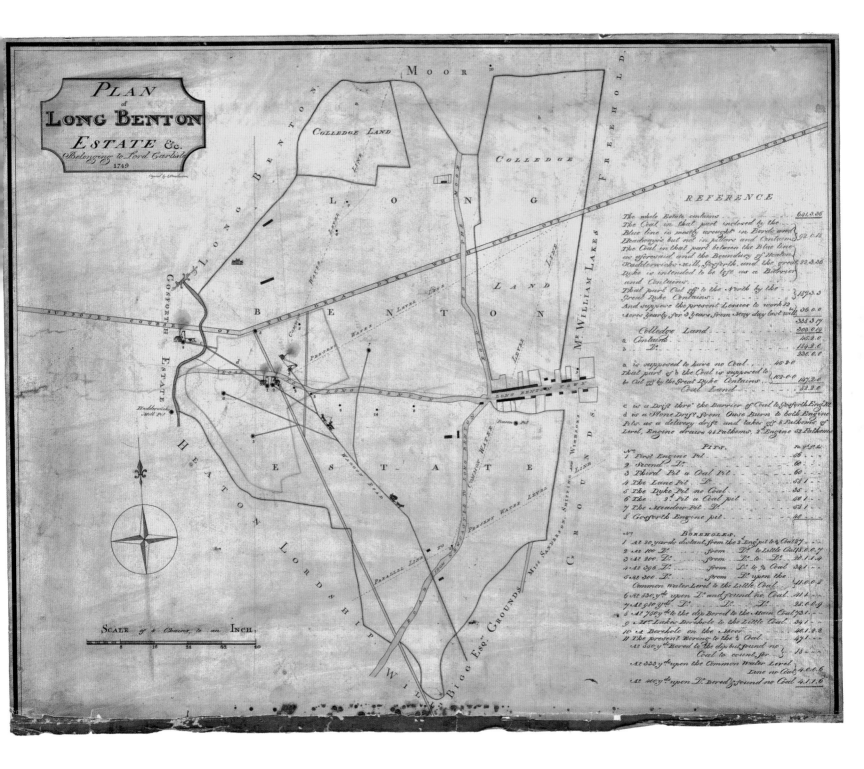

LEFT: J.T.W. Bell, detail of *Plan of part of the Newcastle Coal District* (1847) [ROB] Note site of 'intended dock' and 'intended dock railway'.

to depend on a more widespread use of metal rails. With the expansion of the latter, three iconic Tyneside engineers sought to introduce steam powered locomotives – William Chapman (a canal engineer who had been converted to the potential of railways) at the Heaton Colliery Railway, William Hedley at the Wylam waggonway, and George Stephenson at West Moor in Killingworth. Thus, while the earliest breakthroughs in the development of steam traction occurred elsewhere, it was in the North East and Tyneside that its full potential as vital infrastructure was demonstrated. In 1830, Thomas Oliver described the unique spectacle at Lemington where 'the travelling steam engine dragging from 12 to 20 waggons from Wylam is constantly coming and going'.

Despite some technical problems, these early steam locomotives were generally successful but, for a number of years, some lines continued to use horse traction. Intense rivalry between coal owners led to a complex network of lines being developed. For example, in 1840 the Seghill Colliery Company stopped sharing the Cramlington waggonway and built its own line to the Tyne. In August 1841 this line was also opened for passengers and became the nucleus of the Blyth and Tyne Railway. This heralded the building of mixed-use lines around Tyneside. But, as a steam locomotive network developed, waggonways were either abandoned or re-routed to connect with the new railways at their nearest point. By 1850 all wooden rails had disappeared and by 1880 all the old horse-drawn railways had converted to steam traction.

Similar processes occurred on the south bank of the Tyne, best exemplified by the growth of the Brandling Junction Railway, built in stages from Gateshead eastwards to South Shields and Wearmouth from 1839. This had taken over the old Tanfield waggonway of 1725, modernised and connected to the Newcastle–Carlisle railway. Originally intended mainly as a mineral line, the Brandling Junction attracted passengers. The proprietors also operated an omnibus service in South Shields running from a terminus located at Laygate to the Market Place and then on to the Steam Ferry landing, from where boats ran to North Shields and along the river to Newcastle.

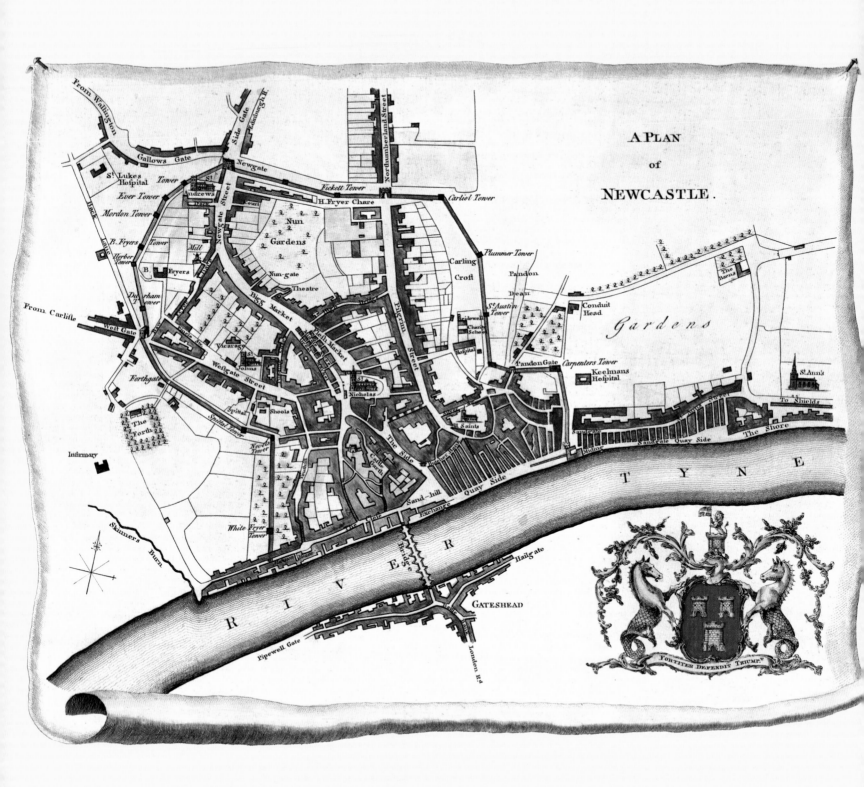

A PLAN
of
NEWCASTLE.

From Wallington
Gallows Gate
Side Gate
Edinburgh R.
Newgate
St Lukes Hospital
Tower
St Andrews
Fickett Tower
Northumberland Street
Carliol Tower
Ever Tower
H. Fryer Chare
Mordon Tower
Court
Nun Gardens
B. Fryers
Tower
Mill
Plummer Tower
Carling Croft
Pandon Dean
Herbert Tower
B. Fryers
Nun-gate
Theatre
Bigg Market
Pilgrim Street
St Austen Tower
Conduit Head
Gardens
From Carlisle
Durham Tower
West Gate
Flesh Market
Bridewell
Charly School
Hospital
Vicarage
Eagle R.
St Johns
Forthgate
Westgate Street
St Nicholas
Pandon Gate
Carpenters Tower
Keelmans Hospital
St Ann's
Spital
Shools
All Saints
To Shields
The Forth
Spittal Tower
Newde Tower
The Side
Sandgate
Quay Side
The Shore
Infirmary
Cowgate
Castle Garth
Butcher Bank
Mining
Sandgate
Quay Side
White Fryer Tower
Sand-hill
Quay Side
Skinners Burn
Manor House
The Gate
Exchange
R I V E R   T Y N E
Bridge
Hailgate
GATESHEAD
Pipewell Gate
London Rd

FORTITER DEFENDIT TRIUMPH.

1769

Newcastle's chares

> Most of the houses not actually fronting the Quayside or the Butcher bank had, towards the close of the last century, been abandoned to less reputable occupiers, who loved darkness rather than light.
>
> F.W. Dendy, *Archaeologia Aeliana*, Series 2, vol. 18, 1896

One of the most striking aspects of the townscape of Newcastle was the plethora of streets that climbed precipitously up the sides of the gorge through which the Tyne flows. Some of the remaining nineteenth-century streets of Elswick to the west and Byker to the east still clamber vertically from the river up the steep slopes to the upper plateau. It is surprising that more of the residential streets were not laid out east–west along the contours of the steep banks. In the central part of Newcastle's waterfront these banks are, in fact, a cliff edge formed by the former course of the river around 70 metres to the north of the present Quayside area and indicated by a steep fall of six to seven metres down to the latter. Archaeological evidence strongly suggests that the current Quayside area is reclaimed land, created mainly by human activity.

The most dramatic of these precipitous streets were the 20 or so chares that lay within the medieval walls on the eastern end of the Quayside. They are clearly shown on Speed's map of 1610 and on the numerous maps that subsequently drew on Speed. The example shown here is the inset drawn on the county map of 1769 by Andrew Armstrong & Son. The map illustrates that the whole of the river frontage had three distinct zones. To the west, linear development took place aligned west–east along the Close. But here there was no public quay and dwellings were built with private wharfs between the river and this west–east route. The central section, lying between the

---

OPPOSITE: Andrew Armstrong & Son, *A Plan of Newcastle upon Tyne* (1769) [AUTH]
Inset from Armstrong's *Map of the County of Northumberland.*

former inlets of the Lort and Pandon Burns, contains the key node of the Sandhill and, not surprisingly therefore, some of the most intense development – 'said to have been the most closely built piece of land in the kingdom' according to R.J. Charleton in his *History of Newcastle upon Tyne* (1885). Development here was aligned north–south, forming the characteristic chares described by Dendy above. To the east, the third zone developed outside the original walls, with the Sandgate providing a central spine for numerous alleyways branching south down to the river or north to the former cliff edge.

Excavations in the central area have demonstrated the process by which this distinctive urban form emerged in the latter two zones. The narrow, linear character and north–south orientation of the chares is due to the creation of a series of piers with docking areas in between, roughly at right angles to the river. These originated mainly in the thirteenth century. Reclamation took the form of revetments, usually made mainly from wattle and then substantial dumping of material behind these. A major source for the latter was ballast material, including imported stone and sand. This created a series of individually owned quays which subsequently guided the orientation of residential and commercial building. The local historian Eneas Mackenzie notes that a reference to Broad Chare can be found as early as 1390, suggesting that the transition from pier and associated docking area to a built up 'chare' or narrow street could have taken place within two centuries.

Armstrong's plan shows the outline of the chares quite dramatically. What is less evident from this and other maps is the steepness of the banks up which many of the chares climbed. Given the topography, most of the chares incorporated numerous steps, were very narrow and passable only on foot. The only one wide enough to allow carts to pass was the aptly named Broad Chare, one of the few remaining named chares.

'Chare' is a word that local historian John Trotter Brockett claimed was peculiar to Tyneside, defining it as 'A narrow street, lane or alley' and suggesting that the word was derived from the Saxon *cyrran*, meaning to turn, the implication being that a 'chare' was a narrower turning from some superior street. Not all chares were on the riverside. Indeed, within the main body of the town there is also Pudding Chare linking the Bigg Market and Westgate Road. It is on relatively flat land but is a narrow slightly winding lane, hence justifying the appellation.

Many chares simply took the name of the principal owner of property. For example, Mackenzie notes that Fenwick's Entry was named after its owner, the alderman Cuthbert Fenwick, who lived in the upper part of the chare. Others described physical characteristics. A stout person would find it inconvenient to squeeze through the upper part of the Dark Chare where the houses on either side nearly touched each other. A third group were named after commodities, for example Peppercorn Chare.

In their early years, as indicated by Dendy, the chares contained some of the finest houses in the town and accommodated opulent merchants, especially those in the coal trade. However, over time the density of the dwellings, their inaccessibility and their increasing dilapidation saw the area become the most squalid and deprived part of the town. They appear also to have included more than a small proportion of houses of ill-repute. The correspondence between John Brand – who published a history of the town and was secretary to the London-based Society of Antiquaries – and the cartographer Ralph Beilby, who engraved the plan of Newcastle for Brand's history, gives some indication of this. In 1788, having rather imperiously ordered Beilby to check on details of the chares, Brand wrote, in self-satisfied terms:

> I am truly sorry that you have had occasion to visit those dark and suspicious lanes, and be thrown in the way of the very dangerous, though not very tempting, females that inhabit them. I cannot help, however, congratulating myself on the very narrow escape I have had, for I had intended to examine them very early in the morning when I was at Newcastle, when if I had been seen either going in or coming out of one of them, my character would have been irretrievably gone.

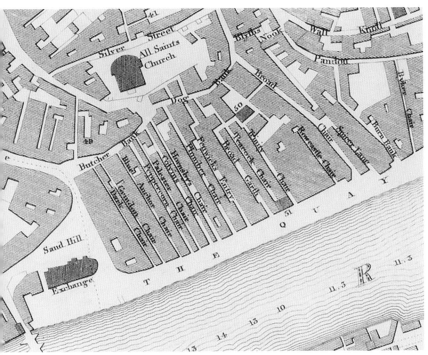

John Wood, detail of *Plan of Newcastle upon Tyne and Gateshead* (1827) [ROB] Number 49 on the map is the Nag's Head Inn, reputed to be the official residence of the mayor and where James VI stayed on his journey south to take up the throne of England.

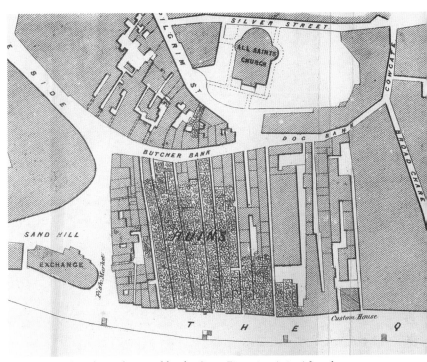

John Rewcastle, *Chares destroyed by the Great Fire, 1854* (1855) [ROB] James Rewcastle ('J.R.') was a printer with a bookshop in Dean Street and the first secretary of the Newcastle Temperance movement.

One of the most detailed cartographic depictions of the chares is by John Wood, whose plan of the town in 1827 shows and names the chares (a total of 18 on this extract) in some detail. Given the high density of development, it is not surprising that his map makes no attempt to distinguish individual buildings within each chare, but it does show the extreme narrowness of many of them. The two important 'public' buildings that he identifies within the area are the Custom House (51) fronting onto the Quayside, and Trinity House (50) at the top of Trinity Chare. Like most other maps of the area, Wood is not able to show the topography of the steep slopes, but one of the fascinating aspects of his map is that, like Corbridge, he consistently spells 'chare' as 'chair', which suggests that, in the absence of anything like street name plates, he simply asked local people the names of the lanes that he mapped and, not being a Newcastle man (he was born close to Barnard Castle and subsequently worked from Edinburgh), he misinterpreted the term. The error does, however, reaffirm the fact that Wood did personally survey the town, which challenges the attempt by Thomas Oliver, his cartographic contemporary, to ignore Wood's work.

The chares suffered heavily from the Great Fire of 1854. The map in Rewcastle's book of the same year shows that one-third of them in the west were destroyed, lying as they did close to the river where much of the conflagration was concentrated. Nevertheless, even in the 1950s it was still possible to clamber down the steps of the remnants of a couple of the chares, and today the names of Pudding Chare and Broad Chare remain as reminders of this fascinating and peculiarly Tyneside phenomenon.

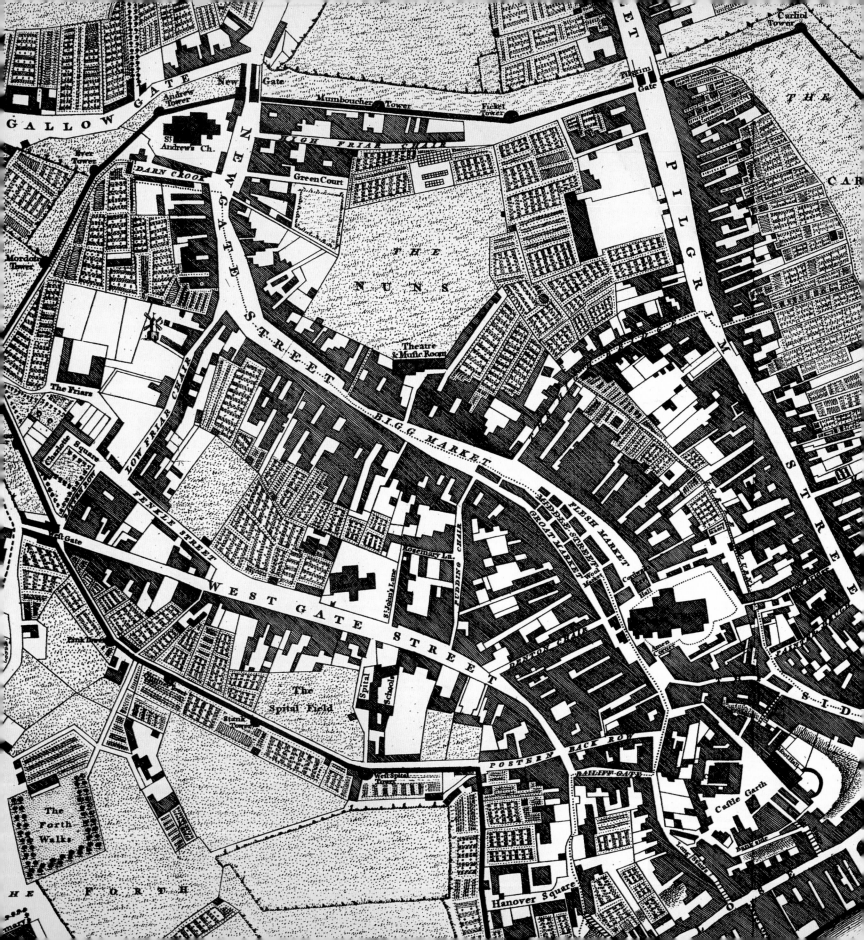

1770

Hutton and Fryer's map

Newcastle by 1770 stood ready for sweeping change, spreading along the river edge, beyond the walls, into the Forth, expanding and developing without order or control.

I. Ayris, *A City of Palaces*, 1999

The finest eighteenth-century plan of Newcastle was this one, surveyed in 1770 and published by Dr Charles Hutton in 1772. Hutton had been born in 1737 in Newcastle's Percy Street and became a distinguished mathematician. He initially worked in the coal mines but, having damaged an arm, was sent to a school in Jesmond. It proved a happy accident because he showed remarkable talent: he became one of the most successful mathematics teachers in the North East and published his first book on arithmetic in 1764. He left Newcastle in 1773 when he was appointed to the Chair of Mathematics at the Royal Military Academy in London, where he played a small role in the origins of the Ordnance Survey and his career flourished with a long list of influential mathematics books. In 1770, while still in Newcastle, he was asked by the mayor and Corporation to prepare a plan of the town.

There are doubts about whether Hutton himself did the surveying. The local surveyor John Fryer was certainly involved. Eneas Mackenzie makes rather contradictory suggestions about the authorship of the plan, saying at one point that Fryer 'measured and drew' the plan, yet at another that Fryer was 'mainly engaged in reducing the large map to the size at which it was engraved'. A neutral observer would find it difficult to imagine that the survey was done by Hutton. In the early 1770s he was running his 'writing and mathematical school', and also writing books – for example, prompted by the 1771 flood, he published *The Principles of Bridges* in 1772.

OPPOSITE: Charles Hutton and John Fryer, detail of *A Plan of Newcastle upon Tyne and Gateshead* (1770) [ROB]
John Scott, later Lord Eldon and Lord High Chancellor of Britain, was a pupil at Hutton's school in Newcastle in 1760.

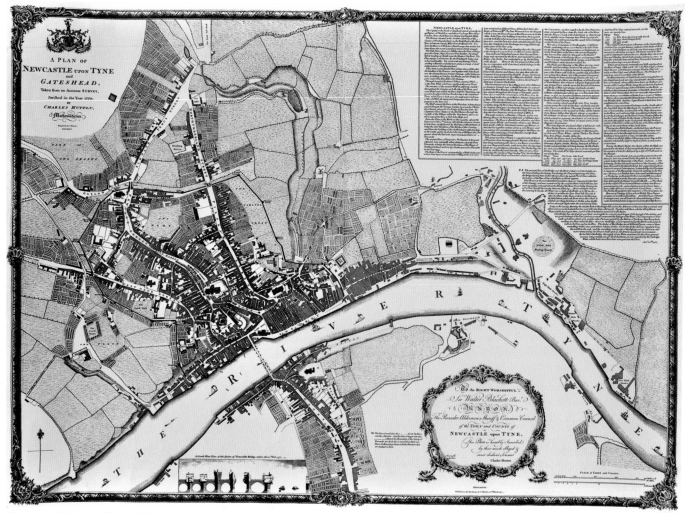

Hutton and Fryer, *A Plan of Newcastle upon Tyne and Gateshead* (1770) [ROB] Hutton left a detailed account of Newcastle's 69 boundary stones, set up in 1648.

How Hutton could have combined all of this with a detailed, time-consuming survey of a major town beggars belief. It seems much more likely that the plan was surveyed by Fryer, despite the map making no mention of his involvement. Fryer was a pupil and assistant to Hutton in the early 1770s and became an accomplished surveyor, producing a number of fine maps including accurate surveys of the Tyne. Fryer's major achievement was the large-scale map of Northumberland which he produced with his sons in 1820 and which, reflecting his

mathematical training, was the first map of the county to show accurate positioning of lines of longitude.

The 1770 Newcastle plan is considerably more detailed than Corbridge's. It depicts public buildings individually and covers a much larger extent of the town environs, as well as much of Gateshead, which now spread along the river front, up Bottle Bank and along the main road to the south. The eastern environs of Newcastle, shown on the full map, indicate the importance of the glass industry, with the 'High', 'Middle' and 'Low' glass works to the east of the Ouseburn. Indeed, the long descriptive account of the town notes the importance of glass as well as such other exports as lead, salt and salmon. The account of the local economy, however, lays most stress on the scale and significance of coal exports drawn from over 20 collieries and employing 'some 150 keels . . . to convey ye coals to the ships at Shields and the other places on the river'. Curiously, however, there is only a single coal staithe marked – and even more curiously it is on the Gateshead, rather than the Newcastle, side of the river (bottom left of the map).

The map shows that, despite the growth of major industries, the town was still predominantly contained within its walls. Many of the linear burgage plots, running back from the main street frontages almost at right angles, show substantial 'in-filling' and increased building coverage compared to the earlier Corbridge map. Nevertheless, the map illustrates some of the significant changes that were harbingers of the future. Hanover Square now sat proudly in the south-west corner of the plateau in an area that was to attract substantial new houses, while Charlotte Square had been built close to the West Gate. However, suburban growth remained very limited.

Unlike many previous maps, the plan attempts to depict the town's topography by using hachures. This is particularly intriguing as Hutton, now Professor of Mathematics at Woolwich, was deeply involved with the group that became the basis of the Ordnance Survey, examining scientific methods of surveying and representing the landscape. The Perthshire mountain of Schiehallion was the 'laboratory' for much of their experimentation and, in this work around 1778, Hutton introduced the technique of using contours rather than hachuring to represent the mountain's relief. Contours were, of course, to become a defining feature of Ordnance Survey maps. But on the Newcastle map of 1770 it was hachuring that was chosen to show the steep climb from the river gorge from White Friars and below the castle. Pandon Dene, Ouseburn and Skinner Burn are also shown as deep valleys cutting through the plateau on their way to the Tyne. Lort Burn is also named, although by this stage virtually all of it had been in-filled and built over.

As the survey was made in 1770, it shows the original Tyne bridge with its nine arches, six on the north owned by Newcastle Corporation and three on the south owned by the Bishop of Durham. However, an inset shows the ruins of the bridge left by the damage caused by the Great Flood of 1771. The map includes an account of the flood and the damage in Newcastle. The central arch of the Tyne bridge and two arches on the south were carried away initially, with other arches collapsing later. Shops and houses on the bridge were destroyed, many of their inhabitants drowned and the structure was left in a ruinous state. Inevitably, damage to the lower part of the town was considerable. Merchandise was washed off the quay and numerous keels and ships were swept down the river. Warehouses, shops and houses were comprehensively flooded from the western end of the Close down to Ouseburn. As it was in the Pennines where torrential rains had fallen while Newcastle had had only light rain, its inhabitants had not anticipated the catastrophic flooding which occurred during the night. Contemporary accounts also described dead bodies, uprooted from graveyards, floating down the flooded river.

Preparations for a new bridge began immediately, with a temporary wooden bridge being opened in 1772. Interestingly, the Corporation commissioned Fryer to prepare a survey of the damage, published as *Plan of such parts of Newcastle and Gateshead as lie Contiguous to the River Tyne*. Hutton's book on bridges drew on this, thereby adding credence to the view that Fryer did the work on the 1770 plan and Hutton published it. The replacement stone bridge was eventually opened in 1781, subsequently being widened in the nineteenth century and being demolished only in 1876 when the Swing Bridge was opened.

Another fascinating detail shown on the map is the reference to 'Amen Corner' beside St Nicholas Church. This was where Ralph Beilby ran his engraving business. Fryer and Beilby clearly knew each other, both having worked for Hutton at various stages. The connection opens up the saga of the wider links between them, on the one hand, and Hutton and the historian John Brand, on the other. It is a tale – recounted in a later chapter – that helped to flavour Brand's history of the town in 1789.

PLAN
of the
River Tyne,
from the Bar or Mouth thereof to the Head of
SOUTH SHIELDS.
surveyed under the Inspection of a Committee of
SHIP OWNERS
by John Fryer of
NEWCASTLE.
1787.

Note. The best Water upon the Bar is with the Low Light House in
Clifford's Fort, and the South Light House upon the Bank in a Line.
The dangerous part of the stones is exactly abreast of the Barracks;
and to pass them safe, you must have the Flag-staff between the
Light Houses; for opposite the West end of the Barracks, the navigable
part of the River is no more than 200 Feet in Breadth.

It is also to be observed, that opposite the West end of Clifford's Fort,
there is only 180 Feet in Breadth where Ships can pass with safety.

SCALES

Chains

Yards.

1787

The growth of the Shields

At the assizes of Newcastle in 1279 . . . it was presented by the jury as a trespass, that the Prior of Tynemouth had built a town on the north bank of the water of Tyne, at Seles, and the Prior of Durham another on the South side, 'where no town ought to be'.

G.B. Hodgson, *The Borough of South Shields*, 1903

Fryer's map of the mouth of the Tyne is as valuable as a depiction of the two 'sentinel' towns of North and South Shields as it is of the navigational difficulties that the river mouth presented. At the end of the eighteenth century the two towns were still small settlements mainly clinging to the immediate banks of the river. Earlier they had played roles as defenders of the river mouth – South Shields with its Roman fort and North Shields/Tynemouth with its castle, Spanish Battery and Clifford's Fort. The two towns developed initially with small

fishermen's huts, 'shiels', from which their names derive.

North Shields/Tynemouth had originally developed under the authority of the Prior of Tynemouth who created a fishing port and built wooden quays to unload fish and to ship coal from the priory's collieries. The town grew despite the attempts by the Newcastle Hostmen to forbid its coal exports, developing both as a fishing port and as an exporter of salt. The town's topography is similar to Newcastle's, with steep slopes from the river leading to a plateau some 60 feet above. This long restricted the settlement to the narrow banks of the river, but by the end of the eighteenth century the town had started to develop on the plateau, with the upper and lower parts of the town connected by steep stairs.

One of the first developments was Dockwray Square, built in 1763. Its elegant houses were occupied by wealthy ship owners and businessmen, while the seamen, fishermen and

OPPOSITE: John Fryer, *Plan of the River Tyne: The Shields* (1787) [AUTH] Fryer had already produced a fine plan of the 'Low part of the River Tyne' in 1773, dedicated to the Master Pilots and Seamen of Trinity House. The 1787 map was a new survey.

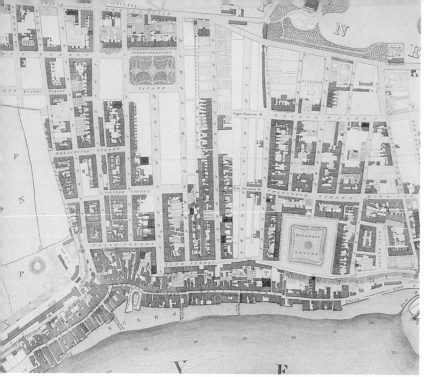

John Rook and Son, *Plan of North Shields and Tynemouth* (1827)
[SANT] Detail of North Shields centre. The plan shows the stark
contrast between the lower town and the upper town.

other working people continued to live in the crowded and
insanitary housing of the lower town. However, the latter was
home to another of the early developments: in 1770 the Market
Place or New Quay, shown by Fryer, was sponsored by the
Duke of Northumberland and developed by the distinguished
Newcastle architect David Stephenson who, inspired by
Edinburgh's New Town, planned its classical Georgian four-
storey buildings. The Market was served by the appropriately
named Duke's Landing and the Northumberland Arms Inn,
the latter constituting the major part of the only one of the
three planned sides of the square that was actually completed.

Fryer's plan shows the significance of shipping to North
Shields, with several roperies stretching along the plateau, as
well as landing quays and the High and Low Lights that helped
to guide ships through the narrow opening of the Tyne. The
earliest of these lights had been built in the seventeenth century,
being rebuilt in 1727 and then replaced by the present-day
lights in 1810.

No large-scale maps of the town were produced until two
detailed plans appeared in 1826 and 1827, drafted by John
Wood and John Rook respectively. Both cartographers sought

sponsorship from the Duke of Northumberland – Rook's map
with its dedication to the Duke, and Wood looking for support
for his 'atlas' of plans of the twelve main towns in Northumberland and Durham. Neither was successful in their quest,
however, and instead over many years the dukes used members
of the local Bell family as surveyors of their various estates.

Rook's is a remarkable plan with details that even include
the types of roofing material of houses. It is an invaluable record
of the expansion of the town onto the upper plateau. By 1827
the stark contrast between the upper and lower town was even
more apparent. Buildings fronting the river along Low Street
were densely packed, while the houses on the plateau were
spacious and laid out along wide, planned streets, nowhere
more so than in Northumberland Square which had become
the location of choice for richer families. Evidence of the Duke
of Northumberland was everywhere apparent, not least in land
ownership. Colonel William Linskill was a local landowner on
a smaller scale, occupying Tynemouth Lodge with its extensive
land and gardens. His son, Captain William, was born in 1807
and became the first mayor of the newly created borough of
Tynemouth in 1849, but the family moved to Warkworth in
1857 and the Tynemouth Lodge estate was sold for building.

Across the river, South Shields is shown by Fryer in 1787
as an equally narrow straggle of buildings. It had developed
partly as a fishing port but more especially through its salt
panning (see 1585). In the eighteenth century it had some 200
salt pans whose fumes generated heavy air pollution. Along
with North Shields, salt production at the mouth of the Tyne
constituted Britain's largest concentration of that activity,
producing over 15 per cent of national output at its early
eighteenth-century peak. Gradually the manufacture of chemicals, glass and shipbuilding began to eclipse salt. As well as
numerous landing quays, Fryer shows three sets of docks and
Thompson's shipyard. He identifies bottle and glass works
which were to become increasingly numerous and important
in the nineteenth century. The topography of the town differed
from North Shields. There were steep banks to the east of the
marketplace and above Herd Sands where Fryer indicates the
Lawe Top with hachuring, but to the west there was a relatively

wide flat area, for example around Mill Dam and to its west where there was quite substantial development. The Market Place is shown with the old 'pepper-pot' eighteenth-century town hall at its heart and adjacent to it the Georgian St Hilda's Church, reputed to stand on the site of Hilda's original seventh-century chapel.

Wood's plan of 1827 shows that most of South Shields still formed an elongated straggle along the riverbank, reflecting the importance of the export trade as well as of shipbuilding and repairing. Along with the various landing places and quays, his map shows numerous docks. Glass-related manufacture is widespread with bottle works, flint glass and other glass manufacturers, the largest of which was Cookson's factory. Indeed, the Cookson family is much in evidence as significant landowners in the town and its environs. The significance of the coal trade is indicated only indirectly with a waggonway leading from the river to ballast hills by the Herd Sands. Collieries had been opened in the town in the early nineteenth century, starting with Templetown and St Hilda's, followed by those in West Harton and Marsden and by the important Westoe Colliery in the early twentieth century.

The concentration of development on the riverbank reflects the essentially industrial nature of the growing town. Nevertheless, Wood's map does show the embryonic shape of its later expansion, with the major shopping area of King Street and, even more significantly, its continuation that eventually led to the sea front, as the seaside tourist industry developed. One of the most interesting details shown by Wood is the proposed chain bridge linking the two sides of the Tyne, between the higher ground of the Lawe and the plateau of North Shields. This would have complemented the ferry crossing which long existed between the two towns, but it was almost 200 years until a tunnel was eventually built, providing an easy crossing for wheeled traffic but lying three miles to the west of the Shields and effectively bypassing the two towns.

John Wood, *Plan of South Shields* (1827) [ROB] Captain Samuel Brown suggested building a suspension bridge between North and South Shields in 1824.

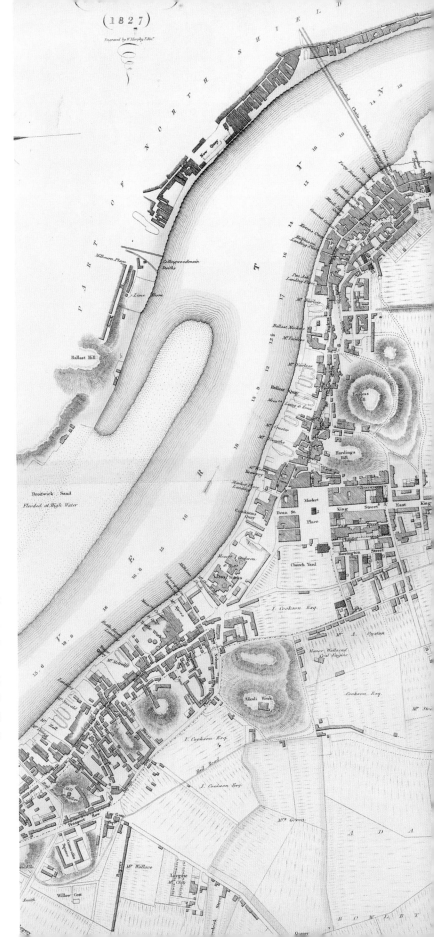

Front of the Publick Baths.

1788

Brand, Beilby and the start of the Georgian makeover

The formation of Dean Street and Mosley Street is the most valuable improvement executed in Newcastle in modern times, but the opportunity is lost of extending them . . .

Eneas Mackenzie,
*A Descriptive Historical Account of the Town
and County of Newcastle upon Tyne*, 1827

At the time of Mackenzie's writing, it may have appeared that his criticism was well-founded, but this chapter signals the beginning of what was to be a remarkable transformation of Newcastle's built environment, the first stages of which actually refuted Mackenzie's pessimism. Three major histories of Newcastle were published in the eighteenth and early nineteenth centuries – Henry Bourne's published posthumously in 1736, John Brand's in 1789 and Eneas Mackenzie's in 1827.

All three men came from relatively humble backgrounds but went on to make successful careers. None flourished more than Brand. He went to the Royal Grammar School, whose formidable headmaster, the Reverend Hugh Moises, recognised his talents and succeeded in getting him sponsored to read theology at Lincoln College, Oxford. Brand attracted numerous backers. He was appointed to curacies at Bolam in Northumberland and at St Andrew's in Newcastle, and in 1774 Matthew Ridley gave him a curacy in Cramlington which he retained for the rest of his life. He moved to London in 1783 where the Duke of Northumberland made him rector of two united parishes in the City of London and appointed him as his personal chaplain and as secretary and librarian at Alnwick. These were heady achievements and may explain Brand's high-handed behaviour to those with whom he subsequently worked. For example, in his editions of *Popular Antiquities* he

---

OPPOSITE: Ralph Beilby, *A Plan of Newcastle upon Tyne and Gateshead* (1788) [AUTH] Beilby was a remarkably talented man,
producing not only this fine plan but also beautifully engraved copper and silver work, and Newcastle's long-stemmed drinking glasses.

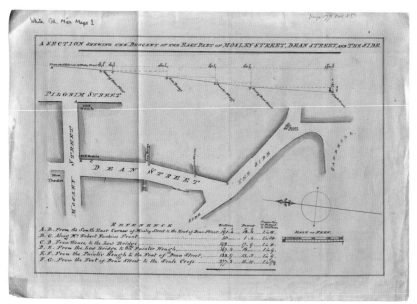

David Stephenson, *Plan and Section of Dean Street from Mosley Street to the Side* (1787–89) [ROB] A good illustration of the problems of Newcastle's topography.

drew heavily on Bourne's earlier work on antiquities to which he added a host of randomly accumulated materials. In his own rather dull history of Newcastle, he drew heavily on Bourne's history, yet unfairly lambasted Bourne's careful scholarship. He was a disorderly manager of his affairs: as resident secretary to the Society of Antiquaries of London, his accounts were found in a hopelessly chaotic state at the time of his death; his magpie-like passion for collecting random miscellanies of archival materials led to rather muddled and rambling publications; he was charged with non-residency of one of his London parsonages since he chose to live in rooms of the Antiquarian Society in Somerset Place; and he was frequently summoned to Alnwick by the Duke which led to his neglecting commitments as secretary to the Society.

Despite clearly being a difficult man with whom to work, Brand was able to include in his *History* a fine plan dated 1788 and signed by Ralph Beilby. There is an element of skulduggery surrounding the plan as much of it is basically a direct copy of the Hutton/Fryer plan of 1770, engraved at about half the scale

of the original. The correspondence from Brand to Beilby shows the imperious way in which Brand ordered Beilby to check or alter aspects of the map, for example: 'What you do, do quickly, and oblige me with another proof.' More interestingly, Brand showed an initial unease about copying the 1770 plan without any attribution. In 1784 he wrote: 'that Dr Hutton will call us to account for the liberty taken in copying his plan without permission . . . ' How the question of copyright was settled we do not know, but in June 1787 Brand wrote to Beilby: 'I should think there will be occasion for no other inscription than "a plan of the town of Newcastle-upon-Tyne with the date of the year".' When Brand's *History* was published, Hutton may have had little concern about the use of a map he had published over 15 years earlier, but Brand must have been worried as legislation in the late eighteenth century meant that plagiarising maps became much more likely to lead to prosecution.

Against this murky background, Beilby produced a fascinating map, especially notable for the streets that he added to the 1770 plan. Despite Mackenzie's pessimism, these signified the start of the major Georgian makeover of the town, one in which local architect David Stephenson played a dominant role. The most important change is the appearance of Mosley Street and Dean Street which effectively started Newcastle's reconfiguration. As the town spread onto the upper plateau, the steep ascent and extreme narrowness of the Side proved a considerable challenge to horse-drawn traffic. Moreover, there was virtually no easy east–west throughway on the plateau itself. The two new streets effectively opened up access to the town, as well as bringing early tastes of elegant Georgian buildings. Named after Alderman Edward Mosley, who played a major role in encouraging physical improvements to the town, Mosley Street was built in 1784 by Stephenson and linked Pilgrim Street to the Cloth Market. He designed a handsome new theatre and a post office that sat on the street. In the late nineteenth century it had the distinction of being the world's first street to be lit by incandescent light bulbs designed by the local inventor Joseph Swan. Even more significant was Dean Street, which was started by Stephenson three years later. It basically followed the line of the lower section of the Lort Burn which, before it was filled in, was described as 'a receptacle of all the filth, butchers offals &c. of the neighbourhood'. The filling-in of the Lort had been delayed because of the high cost of replacing the Tyne Bridge, but once completed it provided wider and more direct access up to the new town that was developing on the plateau.

Most notable among other additions that Beilby made to the Hutton/Fryer 1770 plan, although rather strangely not named, was the new Assembly Rooms at the junction of Charlotte (later Fenkle) Street and Westgate Street, built by William Newton between 1774 and 1776, and financed by public subscription. He also anticipated the replacement of the old All Saints Church, which was falling into decay and demolished in 1786, by a new Georgian church that stood proudly on the summit of the steep banks above the Tyne. The initial design was Stephenson's, but only part of his proposals materialised because his grand Ionic colonnade proved too expensive. The timing was unfortunate for Beilby because the building was only consecrated in 1789, so the plan inevitably gives an inaccurate outline of the completed church, although the uncertainty is perhaps indicated by the dotted lines in front of the building.

Another addition to the earlier plan is the appearance of Saville Row, lying off Northumberland Street. Its row of large elegant Georgian houses was the axis from which more streets were to be built. A further addition to the town and subsequently giving its name to Bath Lane was the 'Publick Baths', displayed at the bottom of the map. Built in 1781 and commissioned by several members of the medical profession, the baths had a large external swimming pool and separate cold plunge pools for ladies and gentlemen. However, the baths were not really 'public', as entrance was for subscribers only. The water supply was taken from the Skinner Burn but was later disrupted by the sinking of a nearby mineshaft. Beilby's map also draws attention to another infrastructural feature – the generous provision of large areas of open public space and gardens – although the pressures of nineteenth-century urban growth were eventually to cause the disappearance of most of these.

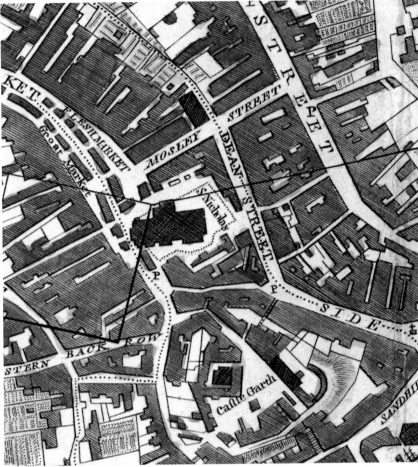

Beilby (1788) [AUTH] Detail of Mosley Street and Dean Street area. The course of the Lort Burn is shown by the dotted line in Dean Street, passing to the east of the rectangular 'New Theatre'. Collingwood Street is not yet built.

Beilby was clearly a distinguished engraver. In his business at Amen Corner, he took on Thomas Bewick as an apprentice in 1767 and made him a partner in 1777. Beilby engraved on glass and copper, while Bewick became celebrated for his wood engravings – a fame which ultimately overshadowed Beilby's achievements. The partnership produced such renowned works as A General History of Quadrupeds, which Beilby wrote and Bewick illustrated. However, disagreement over a later edition led to the dissolution of their partnership in 1798.

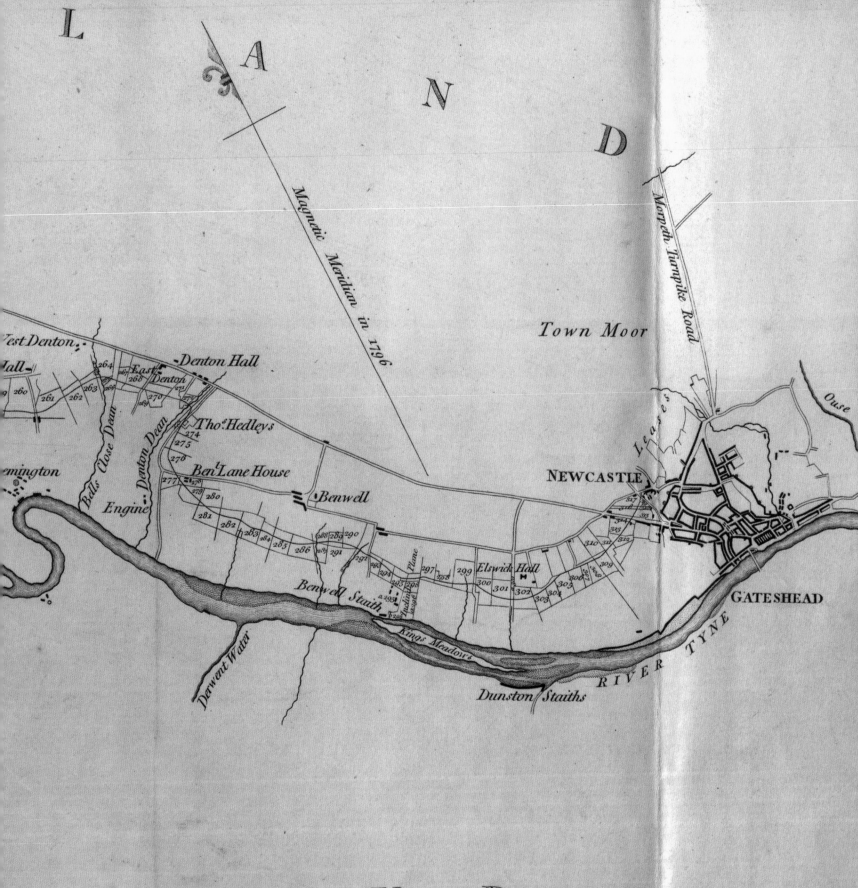

L A N D

Magnetic Meridian in 1796

Morpeth Turnpike Road

Town Moor

West Denton
Hall
Denton Hall
264
267 East
266 268 Denton
263
260 261 262 270 271
272

Tho:Hedley's
274
275
276
Ben.t Lane House
277
279 280
281 282
283 284
285 286
288 289 290
287 291
292
294
295
Plane
297
299 *Elswick Hall*
300 301 302
303 304
305 307
309
306
308
310 311 312
313
314
305

NEWCASTLE

Leases

Ouse

emington
Bell's Close Dean
Denton Dean
Engine
Benwell

Benwell Staith

Derwent Water
Kings Meadows
Dunston Staiths

R I V E R   T Y N E

GATESHEAD

O F   D U R H

1796

A sea-to-sea canal: misplaced ambitions

The failure of the canal meant the triumph of the railway.

W.W. Tomlinson, The North Eastern Railway: Its Rise and Development, 1967

The national canal mania of the late eighteenth century led to a flurry of investment in 'navigation development' in the 1780s. Tyneside was caught up in this, even though cold appraisal suggests how inappropriate canal building was for the North East. The principal market for its chief export of coal was London and the ports of the east coast, and the lower Tyne gave the region all that it needed to service the products of the region. Yet this did not stop numerous proposals to build canals along the east–west line linking the Tyne and Solway.

As early as 1709 an application was made to Parliament to make the Tyne navigable from Newburn to Hexham, but it was not until almost the end of that century that the first serious proposals for a canal appeared. This was based on a 'superficial' survey by Ralph Dodd in 1794, with his preferred line being on the south side of the River Tyne. It was decided to seek a second opinion, and the famous civil engineer William Chapman was requested to investigate. Chapman published a rival scheme which favoured the river's northern bank, leaving the Tyne at Ouseburn, sweeping north of the city and westwards via Elswick to Corbridge. His main reasons for leaving the Tyne as far downstream as the Ouseburn was to avoid both the shoals of gravel and stones deposited by frequent floods and the congestion of relatively small boats in the upper Tyne, as well as the fact that, after the initial rise at the Ouseburn, the canal could proceed at a level trajectory with no need for further locks. The map shown here is a modified version with the canal starting in Newcastle to the west of

OPPOSITE: William Chapman, Detail of *Plan of the proposed navigation from Newcastle upon Tyne to Haydon Bridge* (1796) [SANT] Chapman's detailed, section-by-section survey is self-evident.

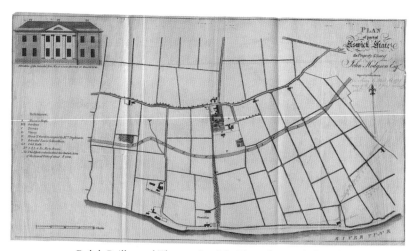

Ralph Beilby and Thomas Bewick, *Plan of part of Elswick Estate showing line of proposed canal* (c.1795) [ROB] It is not clear why Beilby and Bewick produced this map; both are on record as favouring the southern route for the canal.

Gallowgate. Chapman's proposed first stage was from Newcastle to Haydon Bridge and was costed at £3,737 per mile. The route would involve the purchase of land from several large landowners, such as the Elswick estate shown here on a map engraved by Thomas Bewick in one of his rare excursions into cartography.

Such was the enthusiastic atmosphere at this time that an even more ambitious scheme was proposed by Jonathan Thompson, whose preferred point of departure from the river was at North Shields, crossing the North Tyne at Humshaugh. Thompson appears to have been carried away even further as he suggested a branch to Morpeth, Hepscott and Blyth with the possibility of extending even to Berwick. The general enthusiasm for canal development was prompted by a naïve belief that it would automatically boost agriculture, industry and commerce. It seems to have outweighed any logical analysis of the prospects for sustainable trade.

Meanwhile, the two main proponents of a sea-to-sea route had become increasingly confrontational. Ralph Dodd elaborated his initial proposal for the south bank of the Tyne, and in August 1795 Chapman had published a further report

assessing the relative merits of the north and south routes. He acknowledged the greater mineral exploitation potential of the southern route but argued that this was outweighed by the engineering challenges posed by the more difficult topography. In September 1795, a meeting of the subscribers determined that three civil engineers, rather strangely including Chapman, should report further on the northern line. Chapman's northern route was supported by William Jessop, a nationally leading engineer, with the proviso that the canal should start and end to the west of the Tyne Bridge due to the cost of land purchase nearer the city centre. In 1796, Chapman proposed a truncated scheme for a canal on the north bank from Newcastle to Haydon Bridge, shown on the first map. But almost at the same time John Sutcliffe, a notoriously egotistical Yorkshire engineer, had been approached by a group supporting the southern route. However, he proposed a different line from Dodd and dismissed his cost estimates, observing that the stretch from the Tyne to Hexham would cost nearly three times Dodd's estimate. He also sneered that Chapman must have conducted his survey 'from a post-chaise' and that both Chapman and Jessop had grossly underestimated the cost of their northern routes.

In February 1797, Robert Whitworth became yet another appointee to evaluate both schemes. Although departing from some details, Whitworth concluded that the southern line as amended in Sutcliffe's report was preferable, but his refusal to totally condemn Chapman's proposals caused Sutcliffe to refer to him as 'our blundering friend'. However, on 25 March 1797, the Bill for Chapman's Newcastle to Haydon Bridge canal on the north side of the Tyne was considered by Parliament, but the promoters of the southern line led by Sutcliffe and four landowners on the northern route objected successfully. Then the Napoleonic Wars effectively called a halt to the various proposals. In addition, there was a general lack of support for the project at the western end where Carlisle's textile interests gave greater backing to access to the Irish Sea, Liverpool and ports in the west of Scotland.

Activity in the east lay in abeyance until 1810 when a group of interested parties announced their intention to apply for a

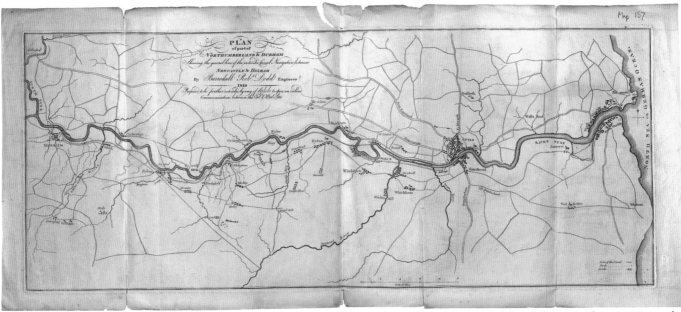

Barrodall Robert Dodd, *Plan of part of Northumberland & Durham shewing the general line of the intended Canal Navigation between Newcastle and Hexham* (1810) [SANT] In contrast to Chapman, Dodd's survey appears to have been much more generalised.

Bill for a canal from Newcastle to Haydon Bridge. This again pitched the Dodd family (this time Ralph's son Barrodall Robert Dodd who retained his father's preference for the south bank) against William Chapman. Dodd's proposed canal began at Redheugh and proceeded to Stella where it met a spur originating at Lemington and bypassing the Tyne's large meander to the south. The continued route on the south bank involved the construction of several groups of locks. As the debate over the route continued, however, it gradually became clear that the whole idea of a canal linking the Tyne and the Solway was probably redundant anyway. A potentially speedier mode of transport was starting to prove itself. Indeed, in May 1824, to the consternation of many, William Chapman made public his preference for a railway. At several public meetings held in Newcastle, testy exchanges took place with 'Willy' Armstrong, father of the future Sir William, who was the leader of a pro-canal majority and celebrated as such in the ballad *The Wonderful Gutter*. This was despite Chapman's carefully calculated evidence that the canal would cost over £500,000 more than a railway. Nevertheless, this was dismissed as being 'partial and unsatisfactory'. Thomas Telford was employed to adjudicate but had to abandon his role due to pressure of other work. He was replaced by a representative of yet another

father-and-son dynasty of engineers, Josias Jessop, the son of William Jessop who had earlier supported Chapman's north bank route. Meanwhile, Chapman had been focusing his efforts on technical innovations at the family rope-making business at Willington on Tyne, observing somewhat resignedly that 'canal business is rather bad'. Jessop's recommendation was unequivocally in favour of a railway and this was finally accepted. Construction of the Newcastle and Carlisle Railway began, and the line was opened in sections, eventually completed by 1839. Much of the route followed Chapman's early survey.

As an intriguing footnote, a proposal came in 1883 from Mr J. Watt Sandeman, a Newcastle-based consulting engineer, for a 'sea-to-sea' canal which proposed not a narrowboat canal but a ship canal. Sandeman argued that benefits would accrue to small towns and their immediate environs along the route, Hexham and Brampton in particular. He also emphasised, rather spuriously, the military advantages of a canal. It would have involved a massive amount of civil engineering and the cost was estimated at a minimum of £44 million. Unsurprisingly, it was never built although the idea was somewhat optimistically revived by officials of Cumbria County Council in October 1996.

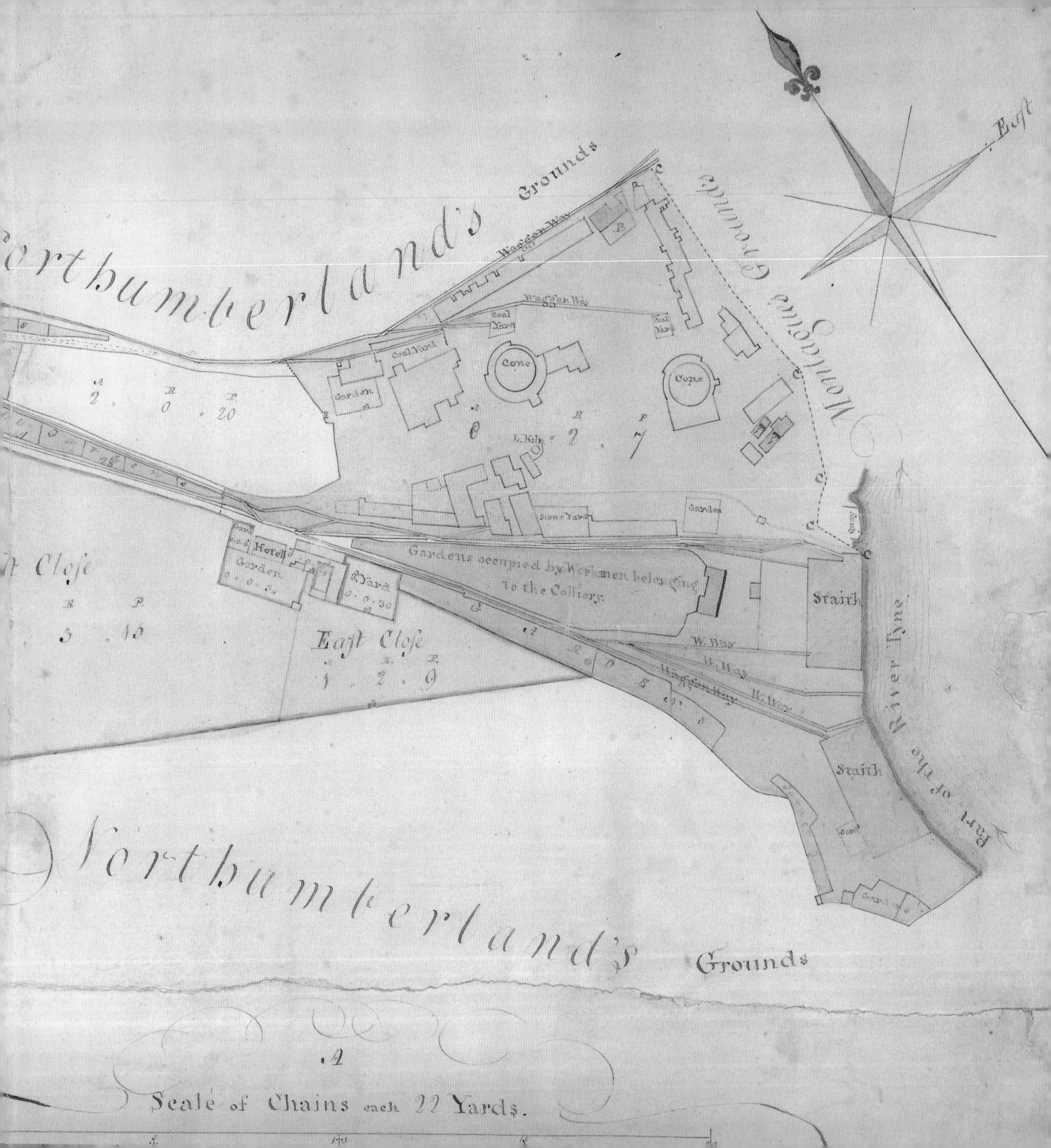

Grounds

*North**umberland's*

Waggon Way

Woodgates Grounds

East

P

Coal Yard

Coal Yard

Coal Yard

Waggon Way

Garden

Cone

Cone

A
2 . 0 . 20

L. Kiln

R P
2 . 7

Close

Coal
Garden

Hotel

Garden
0 . 0 . 24

Stone Yard

Garden

R P
3 . 15

S. Yard
0 . 6 . 30

Gardens occupied by Workmen belonging
to the Colliery.

Staith

River Tyne

East Close

W. Way

W. Way

Waggon Way Hatton

Staith

Coal

Part of the

*North**umberland's*   *Grounds*

.A

Scale of Chains each 22 Yards.

1802

Lost industries of Tyneside

By the end of the Napoleonic War the Tyne had begun to wear [a] black and busy habit . . . From Lemington to Shields its banks were lined not only with collieries but with factories, foundries and forges; with glasshouses, breweries and refineries; with chemical works and shipbuilding yards.

S. Middlebrook, *Newcastle upon Tyne:*
Its Growth and Achievement, 1950

Tyneside may be best known for its coal, ships and heavy engineering, but there were numerous other industrial enterprises that made the area a substantial powerhouse in the nineteenth century. This industrial growth and its associated activities gave rise to a significant increase in the maps and other representations of industrial premises. These reflected new or increased requirements for such things as expansion plans, insurance, the provision of new on-site infrastructure (railways and gas, for example) and marketing. Apart from specialist historians, these maps are probably amongst the least celebrated historical cartography, but they constitute an important genre. We will examine three examples here.

One of the most important industries that subsequently disappeared in Newcastle and Tyneside was glass making. It was hugely important, both in its own right and as a stimulus for other industries. The 1802 plan of Lemington glass works by Thomas Wilkin captures one of the most important firms in a location typical of many industrial sites along the river in the early nineteenth century. It shows the works themselves (outlined in red), opened in 1787, as well as waggonways, coal staithes, some cottages and a brewery. One of the waggonways is the Wylam track upon which coal was brought from five miles to the west and which was so significant in early steam

---

OPPOSITE: Thomas Wilkin, *Lemington Glassworks* (1802) [N.EST] Thomas Wilkin was mainly employed
by the Duke of Northumberland and produced many surveys of the Duke's various properties.

C.T. Maling & Sons Ford Potteries (c.1900) [TWA] The railway in the foreground is the riverside line of the Newcastle–North Shields railway, passing the large Ford B Pottery on its west side before swinging east along the river.

locomotive development by William Hedley. The circular structures are glass cones, built in 1797, one of which survives today as a listed memorial to Tyneside's industrial heritage. A noticeable feature of the map is the provision of gardens, but the key states that these were provided for workers at the nearby colliery. A later map of 1850 by the prolific land surveyor Thomas Bell shows an additional third cone and much intensified development, with many of the gardens built over. But the glass industry in this location was by then on the point of decline.

Newcastle and South Shields both played major roles in glass making. Sir Robert Mansell's site in the Ouseburn Valley in 1617 was the first purpose-built glass works to use coal, and he also drew on cheap supplies of sand brought in as ballast by returning colliers. Other ballast materials such as flint provided the basis for the pottery industry. After Mansell's death in 1652, Huguenot families from Lorraine such as the Tyzacks, Henzells and Tytorys took over the glass works, bringing with them their long experience of Continental glass making. A further glass-making site was in the Close in Newcastle. Here at the end of the seventeenth century an Italian

family, the Dagnias, opened no fewer than four glass works including one that made Tyneside's first flint glass. It was the Dagnia works that produced Newcastle's long-stemmed flint drinking glasses that Ralph and Mary Beilby engraved so beautifully in their business at Amen Corner. Ralph Beilby engraved not only glass but also the copper plates that were used for maps of the town, such as in Brand's *History*. Unsurprisingly, his panorama of Newcastle (*see* 1745) shows smoke belching from the Dagnia works in the Close, which by this time was being abandoned by its richer inhabitants.

In the late eighteenth century, the Dagnia business was bought by John Cookson & Company, which came to dominate Tyneside glass making, especially through its major South Shields works. The firm specialised in crown glass for windows. Along the riverbanks in South Shields there were eight large glass works, and the Cooksons, through a variety of partnerships, were directly or indirectly involved in most of them. By the mid nineteenth century they had sold their business to a partnership that included men such as George Stephenson and George Hudson. It established a new plate-glass business whose works at Mill Dam – together with the Sunderland glassmaker James Hartley – won the prestigious order to manufacture the glass used for the Crystal Palace in the Great Exhibition of 1851.

There was yet more glass making south of the river in Gateshead, where, at the end of the eighteenth century, Charles Attwood formed the Tyne Glass Company and took over an existing glass works at Saltmeadows. In 1772 there were 16 substantial factories in Newcastle, and the combined output of Tyneside glassmakers accounted for no less than 40 per cent of English glass in the early nineteenth century.

As well as its importance as a site for glass, Ouseburn was also a centre for much of Tyneside's manufacture of pottery, aided by the availability of materials derived from ballast that formed the basic raw materials for this industry. Most of the pottery was aimed at a middling market rather than the high-quality products of Staffordshire and Worcester. By the early nineteenth century there were some 20 potteries on Tyneside, most located in Ouseburn where the principal manufacturers were the Maling family, French émigrés who had moved to England in the sixteenth century, set up a pottery in Sunderland in the eighteenth century and moved to Ouseburn in the early nineteenth century. In 1879, the firm expanded to a 14-acre site in St Lawrence, known as the Ford B pottery, just to the east of the Ouseburn Valley, and by 1901 employed some 1,000 workers. The firm developed a core business in the mass production of cheap earthenware containers establishing, for example, a lucrative long-term arrangement to supply jars to Keillers, marmalade makers of Dundee. However, a growing interest in high-quality decorative work allowed the firm to diversify and open London sample rooms, and by the 1920s it had become renowned for the quality of its richly enamelled and gilded designs. The postcard advertisement shown here reflects this 'fashionable' side of the business with not only the size of the business and its railway connections being meant to impress, but also its possession of 'London Sampling Rooms' in Holborn Circus. The Maling potteries survived until the early 1960s and, on closure, had no less than 16,500 designs on their pattern books.

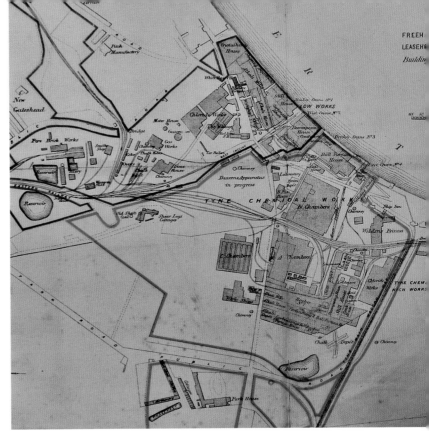

Plan of Tyne Chemical Works in the Borough of Gateshead (1873) [GLH] Known locally as Allhusen's Chemical Works long after ownership had changed.

Alkali was in great demand from the glass industry and provided a stimulus for the growth of numerous chemical manufacturers on Tyneside. The large alkali works at Walker were founded by the Losh family who used brine to manufacture soda. Sulphuric acid was produced at Saltmeadows in Gateshead, and a soap factory was opened further downstream at Friars Goose by Anthony Clapham in 1827. His works contained the tallest chimney on Tyneside (263 feet high), and later manufactured bicarbonate of soda and Epsom salts. A number of soap factories were set up in Ouseburn, but the most significant was the huge works in Newcastle that became Thomas Hedley & Co. in City Road with a site reaching down to the river. This operated for over 150 years and was eventually bought by Proctor & Gamble.

But the linkage between glass manufacture and chemicals is best illustrated by the Allhusen's Chemical Works located at South Shore, Gateshead. The firm had initially been developed by Charles Attwood in the early 1830s for glass manufacture, expanding into soda production in 1835. In 1840 the firm was purchased by Christian Allhusen and it expanded rapidly. By the early 1870s, however, Allhusen's was struggling, and the map of 1873 was produced as a document relating to the sale of the site. The firm was taken over by a public company, the Newcastle Chemical Works, although it continued to be referred to as Allhusen's. A particularly significant feature of the 1873 map is the use of railways within the site to move around large quantities of the bulky and often dangerous materials. The site extended over 130 acres, manufacturing mainly soap and alkali but also pitch and bricks. By 1889, 1,200 men were employed, but by this time, like the Friars Goose Chemical Works, the firm had been bought by the United Alkali Company and the main product was caustic soda.

These chemical works caused much environmental damage, but their sites have now disappeared, most being replaced by open green areas such as East Gateshead Riverside Park and Hebburn Riverside Park.

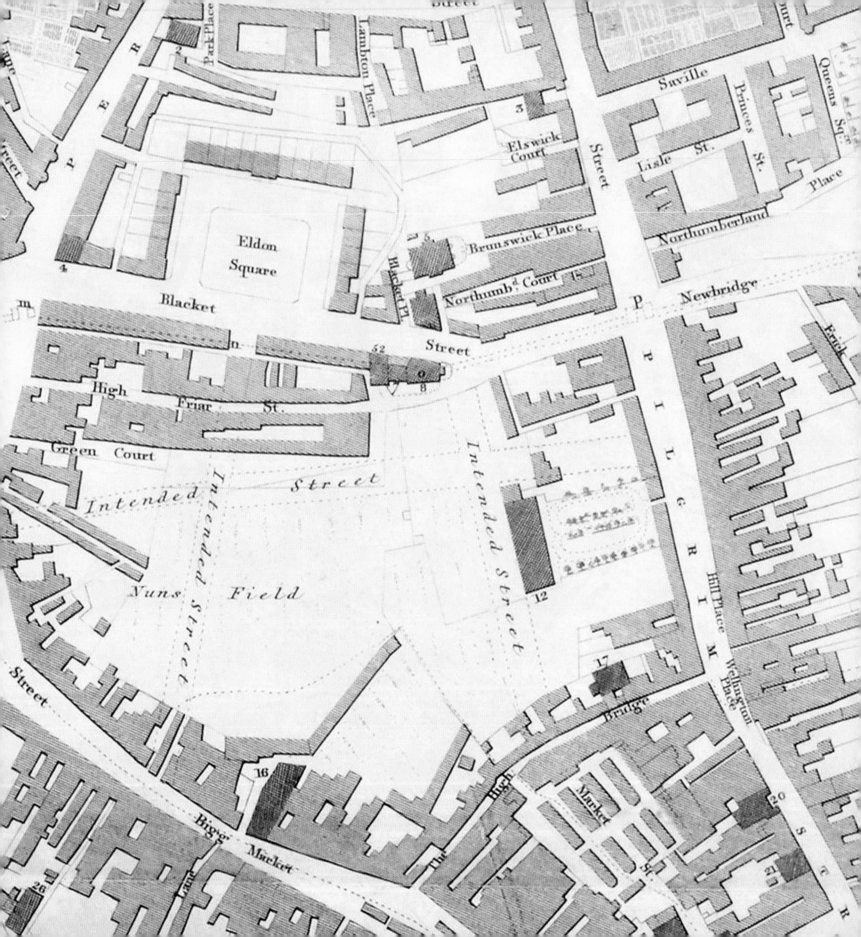

1827

Early Georgian makeover

It must . . . not be forgotten that at that time [1825] nothing yet existed of Grainger Street and Grey Street. The centre of Newcastle except for Mosley Street was still narrow and unplanned. The spaciousness of Eldon Square was a thing quite unheard of.

N. Pevsner, *The Buildings of England*, 1957

The dramatic reconfiguration of Newcastle in the Georgian and early Victorian eras differed from other British towns with classic Georgian architecture. The Georgian squares in London and in Bath were new additions to their peripheries, and Edinburgh's New Town fashioned a wholly new townscape well beyond its medieval core. In Newcastle, by contrast, new building largely replaced the earlier streets and houses as well as developing what had been open land within the town's boundary.

John Wood's plan covers the early years of this transformation that had started in the late eighteenth century. His plan is one of the dozen large-scale plans that he drew of towns in the North East in the 1820s. Of the impressive number of some 150 detailed town plans that he produced across Britain, 76 were of English towns. He lived in Edinburgh and his first forays south into England were to make plans of the border towns of Carlisle and Durham in 1820, and of Berwick in 1822. He followed these with other north-eastern towns: in 1826, Morpeth, Hexham, Sunderland, Darlington, Stockton, Alnwick and North Shields/Tynemouth; and in 1827, Newcastle/Gateshead, South Shields and Barnard Castle. His plan of Newcastle is typical of his distinguished and detailed cartography.

The principal interest in Wood's plan is its depiction of the changing face of Georgian Newcastle. The most dramatic new

OPPOSITE: John Wood, detail of Central Newcastle from *Plan of Newcastle upon Tyne and Gateshead* (1827) [ROB]
The 'intended streets' shown by Wood are essentially those suggested by John Dobson and described by Mackenzie also in 1827.

feature was Eldon Square, designed by the distinguished architect John Dobson and built by the buccaneering developer Richard Grainger from 1824 (which Wood's map correctly identifies in pink as not having been fully completed by 1827). Blackett Street had been built on the southern flank of the square, replacing the line of the northern walls of the town. The earlier roads giving better access to the town had been extended, so that Collingwood Street now continued the line of Mosley Street linking through to Westgate Street. There was now a striking contrast between the old and the new, shown in the juxtaposition between the long narrow medieval burgage plots running from Northumberland Street and, cheek by jowl, the new wide streets built on what had been Carliol Croft. This development expanded the area of fashionable houses built in and around Saville Row, a development which itself now extended further east along Ellison Place. The town's new gaol, another John Dobson design started in 1822, was nearing completion on the southern end of the Croft and, to the east, New Bridge Street had been formed with a bridge crossing the deep ravine of Pandon Dene leading to the substantial new houses of Ridley Villas, built on land owned by Sir Matthew Ridley. New developments had eradicated large sections of the old defensive walls: only the western section and a very small stretch on the east remained. The town had not only burst beyond the bounds of its walls, but within the old intramural area much had been, or was in course of being, replaced.

Wood also shows some of the major new public buildings that had been completed by 1827. Among the grandest, the County Courts, built in Grecian style with imposing pediments and porticos on its north and south sides, opened in 1812 in a strategic site on the plateau next to the remains of the castle and overlooking the Tyne Bridge. But more symbolic were two other buildings, John Dobson's Northern Academy of Arts building (1827) on Blackett Street, shown as number 52 on Wood's map, and the purest Georgian of the new home of the Literary and Philosophical Society on Westgate Street, completed in 1825. The design of the 'Lit & Phil' was by John Green, the local architect who worked with his son, Benjamin, on the plans of a number of railway stations, churches and other

buildings in Newcastle and the North East. Originally founded in 1793, the Society played an influential role in the intellectual life of the town, attracting notable speakers, amassing a superb collection of books for its substantial library and offering accommodation for fellow societies such as the Antiquarian Society of London. Joseph Swan demonstrated his newly invented electric light bulb at the Lit & Phil, and the building became the first public building to be lit by Swan's bulbs. Earlier, in 1815, George Stephenson had made a presentation of his 'Geordie' miners' safety lamp to the Society. Its members played key roles in creating new institutions and societies in the town, laying the foundations for institutions such as the Society of Antiquaries and the Natural History Society.

One of the striking aspects of Wood's plan is its series of 'proposed streets', one continuing the line west from Collingwood Street and another continuing Blackett Street to the west, while on the large open space of Nuns Field there is a grid of proposed streets. They reflect the fact that the town was in the midst of major redevelopment and that numerous proposals were being touted, not all of which would materialise. Surveyors like Wood were, to a large extent, left to guess which to show and which to ignore.

More generally, his plan lists the gates and towers of the town walls. His references to individual buildings include schools, hospitals, gas works, a 'racket' court, theatre, post office, banks, inns, and (as with most of his maps) a bewildering array of churches and chapels of various denominations. He also depicts the formal borders of the town, showing the boundary stones by which it was delineated.

As well as providing a comprehensive naming of streets, he also shows something of the topography of the town, especially the ravine of Pandon Dene which is suggested by his use of shading and hachuring. He even attempts to suggest the steepness of the banks above the gorge of the Tyne, with hachures on the Gateshead side, some shading on the western end of the Newcastle banks and the steps shown on Tuthill and the Long Stairs. River depths are displayed, again suggesting the river's navigational challenges as they are less than five feet deep in places.

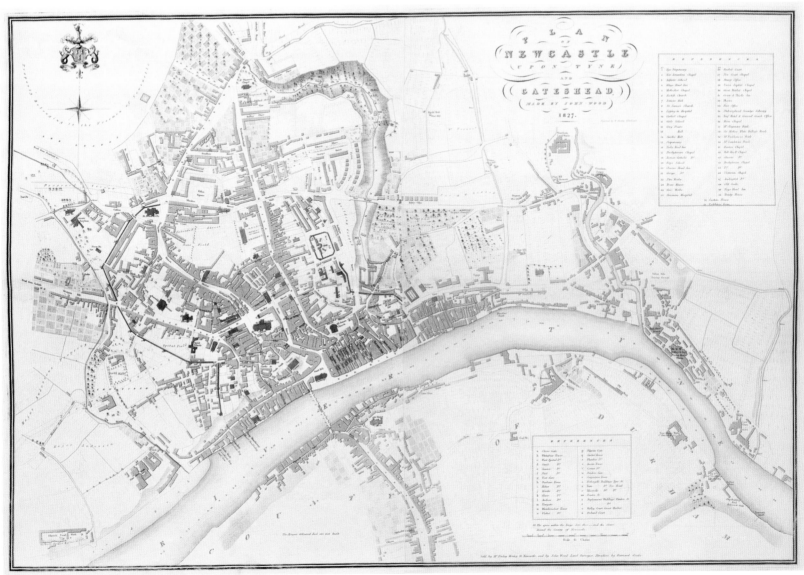

John Wood (1827) [ROB] Wood proposed an elongation of the Collingwood Street axis to the west, towards the Forth.

The links between the river and the growing industrial importance of the town is very evident. From west to east, the banks of the river are shown to have numerous coal staithes, lead works, iron foundries, bottle and glass works, soap works, rope walks and paper factories. Wood's map also shows the rapidly growing industrial activity up the Ouseburn Valley, as well as on both banks of the Tyne downstream from its junction with the Ouseburn. These include lead works, potteries, a tannery and several major glass works – the High Glass Houses, Ridley's Middle Glass Works and the Low Glass Houses – on the Newcastle side, and a foundry, paper manufacturer, glass works and shipbuilding yard on the Gateshead bank.

Plan of the Town & County of
NEWCASTLE UPON TYNE
AND THE BOROUGH OF
GATESHEAD
With their respective Suburbs:
Shewing the Buildings and different Properties contained therein
From an Actual Survey by T. OLIVER, Architect and Surveyor:
Accompanied with a Reference of Proprietors' names.
Public Buildings &c.
1830.

1830

Completing the classical townscape

You walk into what has long been termed the COAL HOLE OF THE NORTH and find yourself in A CITY OF PALACES, a fairyland of newness, brightness and modern elegance.

W. Howitt,
Visits to Remarkable Places, 1842

The series of superb plans of Newcastle produced by Thomas Oliver capture the sequence of the town's later Georgian development. Oliver was both a surveyor and an architect. He had been an assistant to John Dobson but set up as an independent architect in 1821. There appears to have been little love lost between them, and Dobson's daughter published a book praising her father's achievements, largely dismissing Oliver, some of whose work she attributed to her father. Oliver was not a local man, having been born near Jedburgh. His spectacular plan dated 1830 and published in 1831 has been widely lauded as the finest plan of the town ever produced. In *British Town Maps: A History* (2015), Kain and Oliver claim its production was unusual in 'enjoying the active support of the municipal authorities', but (*see* 1723) the Newcastle Common Council had already supported Samuel Buck and James Corbridge in the same year. However, Oliver's map of 1830 was 'so perfectly executed as to invest the name of Thomas Oliver with a reputation that will continue to be associated with the town itself'. It is a huge map, depicting the town in great detail at a scale of 1:2,376. He published a much-reduced version in 1831 (re-issued in 1833) to illustrate his *Picture of Newcastle upon Tyne*. His elegant plans include two commissioned by the Council, one in 1824 showing sewerage lines

OPPOSITE: Thomas Oliver: detail of *Plan of the Town and County of Newcastle upon Tyne & the Borough of Gateshead with their respective suburbs* (1830) [BL] This magnificent plan was accompanied by a detailed book of reference listing the ownership and value of all numbered buildings and plots.

and other public utilities, and a second in 1852 depicting detail of the Town Moor and Castle Leazes including intakes and wayleaves (*see* 1852–53). A further large-scale plan in 1844 was drawn at a scale of 1:6,336. His final general plan, showing the Central Station and the High Level Bridge for the first time, was surveyed in 1849 and updated in 1851.

The large 1830 plan shows that Eldon Square had been completed, and there was further development of areas along Westgate Hill, north of Percy Street, and to the east in Sandyford. Oliver's smaller map of 1833 shows a much larger number of projected streets, compared to Wood in 1827 (*see* 1827). In the flurry of town development, Oliver submitted plans to the Common Council for straightening and improving existing streets and for wholly new streets; as depicted on his 1833 map. However, they did not find favour with the Council who considered the proposals too ambitious and expensive. There is little similarity between the projected roads shown by Wood and by Oliver. For example, one of the proposed roads shown by Wood follows the line that was to become Grey Street immediately to the west of Anderson Place, but it does not appear on the Oliver map. Instead, Oliver proposed an east–west street across the Nuns Area. Other schemes included a western extension of Blackett Street and an entirely new street across the north of the town centre but south of Barras Bridge, cutting across the top of Northumberland Street and Percy Street. In the southern part of the town, Oliver proposed a widening of the street leading from St Nicholas Square directly to the Castle Garth and south of Westgate Street another completely new street leading westwards from the institutional building complex formed by the Lit & Phil and Natural History Society. Oliver later added an intriguing detail to this smaller 1833 version with a projected Tyne bridge to be built above the eighteenth-century one. His inscription reads 'Projected Bridge over the STONE BRIDGE', the latter being the Tyne Bridge.

Oliver's detailed plan of May 1834, which is clearly much more of a working document than his other beautifully executed maps, shows the major developments in the construction of 'Grainger Town' and the creation of Grey Street

(initially called New Dean Street). They were the core of Grainger's design and involved developing the sites of Anderson Place and Nuns Field.

The breakthrough that allowed this to proceed was the death of Major Anderson who had resisted selling Anderson Place. After his death his executors consented to sell the estate and Grainger bought it, together with the adjacent Nuns Field, in total an area of some 12 acres which had long remained virtually open land in the heart of the walled town. It was the site on which Grainger built Clayton Street (whose proposed extension to Fenkle Street is shown by dotted lines) and Grainger Street which, with Nun Street and Nelson Street, formed the large rectangle on which Dobson designed the new indoor Grainger Market, which was opened with great ceremony in 1835 and was praised as being the largest and finest in Europe. Grainger's plan for what became Grey Street entailed the demolitions of the previous Theatre Royal in Mosley Street and of the Butcher Market, shown on the 1834 plan, which had only been open for 30 years. The all-powerful town clerk, and Grainger's unofficial advisor, John Clayton, persuaded the Council to agree to this development since Grainger undertook to build the new indoor market at the heart of Grainger Town, as well as a replacement theatre on Grey Street. As shown on the plan, the original intention was to build it on the west side of Grey Street, but it was eventually built on the eastern side. A further change in the eventual building was the omission of one of the cross streets between what was to become Grey Street, called here 'New Dean Street', and Pilgrim Street.

Completed in 1839, Grey Street was designed by a number of local architects, including Dobson himself, the partners John Wardle and George Walker, and Benjamin Green who was responsible for the new Theatre Royal and Grey's Monument. The wide street climbs up to the monument, which provides a dramatic visual end point, while the bold portico of the theatre provides a fulcrum around which the street gently curves. Few would disagree that Grey Street was the pinnacle of the town's reconstruction.

By the time of Oliver's 1849 map, the Georgian and early

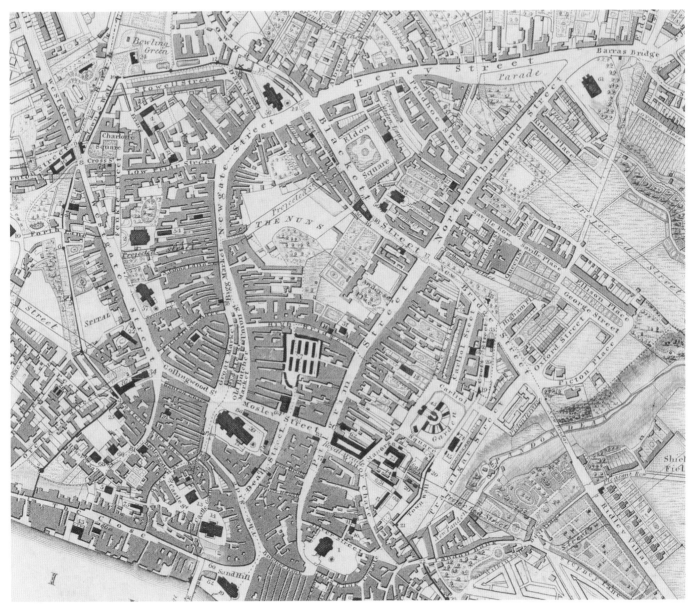

Thomas Oliver, *Plan of Newcastle upon Tyne and the Borough of Gateshead . . . shewing the projected improvements* (1833) [SANT] This 1833
plan is an update of a smaller version of Oliver's 1830 plan and was published in his *New Picture of Newcastle upon Tyne*.

Victorian development was virtually complete. In addition to
the building of Grey Street and Grainger Street, Clayton Street
had been extended to Westgate Street although on a line slightly
to the north-west of that shown on the 1834 plan. The visual
townscape was transformed, and notable new buildings now
graced the skyline. Most were the outcome of Grainger's overall
design and developer expertise, and Dobson's architecture,
along with Benjamin and John Green, John Wardle and George

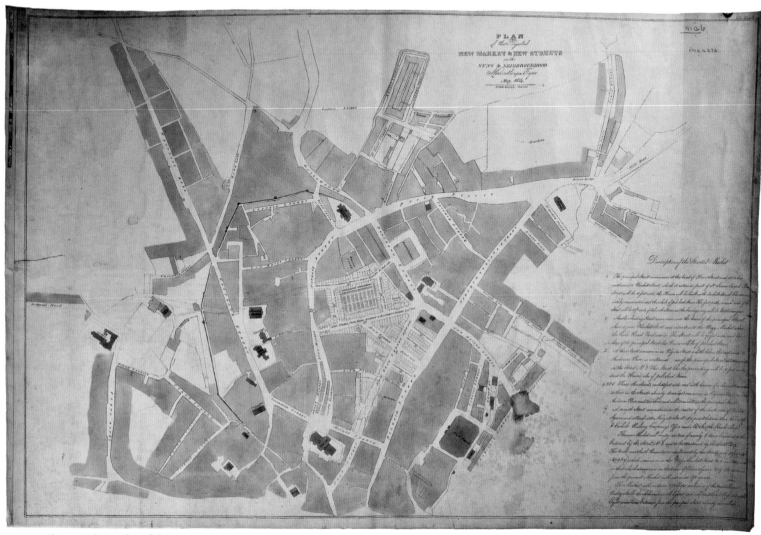

Thomas Oliver, *Plan of the projected new market and new streets in the Nuns and neighbourhood* (1834) [ROB]
Oliver notes that the 'Market will contain 276 shops exclusive of the vegetable and poultry stalls'.

Walker. The Royal Exchange at the bottom of Pilgrim Street was opened in 1832 (called Post Office Arcade, number 54, on the 1849 map), although its off-centre location proved less than ideal to attract custom. The Central Exchange (number 15 on the map) was built in the island formed by Grainger Street, Market Street and Grey Street. Especially prominent was the 'Butcher and Green Market' (later Grainger Market) (numbered 82) and the Theatre Royal (number 74). Leazes

Terrace and Leazes Crescent, north of Percy Street (now outrageously overshadowed by the football ground of St James' Park), were elegant residential additions, called by Oliver 'an ornament to Newcastle, and one of the most pleasant and desirable residences of the town', immodestly ignoring the fact that it was he who was the architect.

The 1849 map also illustrates the influence of railways in re-shaping the southern parts of the core area, especially with the impact of the newly built Central Station and Stephenson's High Level Bridge (*see* 2021b). The Central Station was Dobson's design. He made dramatic use of the awkward site on which it was built by designing a unique curved roof that made innovative use of malleable rolled iron ribs for its three arched roofs and a portico that was added later; less grand than Dobson's original proposal, but nevertheless a striking entry to the town.

As an architect, Thomas Oliver was less celebrated than John Dobson, but nevertheless left a distinguished legacy to the town with his design of Leazes Terrace, and demonstrated his versatility in building for the solicitor and property developer George Tallantire Gibson the artisans' neighbourhood of 'Gibson Town' between New Bridge Street and City Road. This is shown under development on the south-east corner of the 1833 plan and included the innovative, but long-demolished, Victoria Bazaar (intended to function as the 'Grainger Market' for the east end of the town). Some of Oliver's proposed new streets in his map of 1833 demonstrate his attempt to link 'Gibson Town' into the structure of Newcastle with, for example, a street from Stepney Lane cutting across Trafalgar Street to the Royal Arcade and Pilgrim Street. Even more ambitious (and remarkably prescient) was, in effect, a proposed eastern inner ring road running from Stepney Lane and 'Gibson Town' northwards across New Bridge Street and Shieldfield, then curving around the north-east of the town to meet Lovaine Place to the east of the church of St Thomas the Martyr.

Oliver's long series of maps of the town provide a superb portrait of the changing face of Newcastle. Nevertheless, he played second fiddle to Dobson and in his lifetime never achieved the acclaim he deserved. Dobson, a local man born in North Shields, was far more widely lauded and given more commissions. He designed more than 50 churches and probably 100 private houses scattered across the North East. He was elected as the first President of the Northern Architectural Association and there is no question that he was seen as the premier architect in the North East. His neoclassical designs have left a defining mark on the townscape of Newcastle.

Thomas Oliver, detail from *Plan of Newcastle upon Tyne and Gateshead from Actual Survey* (1849) [GLH] Note that Grainger Street originally stopped at the Bigg Market. It was extended to Westgate Road in 1869 to provide better access to the Central Station.

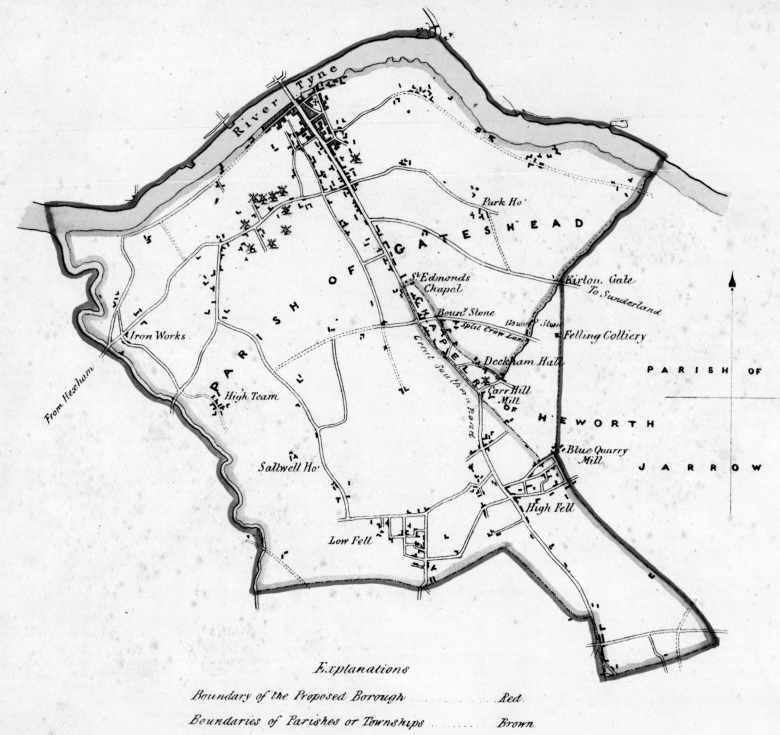

River Tyne

PARISH OF GATESHEAD

Park Ho

Kirton Gate
To Sunderland

St Edmonds Chapel

Bound.y Stone

Boundary Stone

Split Crow Lane

Felling Colliery

PARISH OF

Iron Works

CHAPELRY

Deckham Hall

Carr Hill Mill

From Hexham

High Team

OF

HEWORTH

Blue Quarry Mill

JARROW

Saltwell Ho.

Great Southern Road

High Fell

Low Fell

Explanations

Boundary of the Proposed Borough . . . . Red.

Boundaries of Parishes or Townships . . . . Brown.

Rivers . . . . Blue.

Rob.t K. Dawson

Lieut R.E.

Scale of 2 inches to 1 Mile.

Furlongs 8   7   6   5   4   3   2   1   0   Mile

C. Bradbury Del.t

R. Cartwright Lithog.

1832

Municipal reform

The old borough limits were being overflowed so rapidly that it was impossible to cope with the new conditions. Problems of lighting, paving, drainage, water supply and police became more and more difficult – and costly – to deal with.

S. Middlebrook, *Newcastle upon Tyne: Its Growth and Achievement*, 1950

It was in the fourth decade of the nineteenth century that the first meaningful attempts to respond to the ongoing urban transformation were made, albeit somewhat reluctantly. The Great Reform Act of 1832 has a particular salience for Tyneside and the North East. Charles Grey, 2nd Earl Grey, was the Whig prime minister who came to power in 1830 and helped to push parliamentary reform through a fractious Commons and a hostile House of Lords. Grey came from a long-established Northumberland family based at Howick Hall. Grey's Monument, erected in 1838, stands dramatically at the head of Grey Street, and the inscription on its plinth reads: '[He] . . . was the Minister by whose advice, and under whose guidance, the great measure of parliamentary reform was after an arduous and protracted struggle safely and triumphantly achieved in the year 1832.' However, by modern standards the reform was a distinctly modest step towards universal suffrage. The electorate remained entirely male and, being based fundamentally on the value of property ownership, only increased the electorate from about 400,000 to 650,000, which represented about one in five adult males. By far the greatest extension of representation was among the textile towns of Lancashire and Yorkshire. In contrast, the North East saw

OPPOSITE: Robert K. Dawson, *Gateshead: proposed borough boundary* (1832) [AUTH]
Robert Kearsley Dawson had been an assistant to Thomas Colby in the Trigonometrical
Branch of the Ordnance Survey's triangulation of Scotland.

relatively little change to the numbers of its MPs. Morpeth's existing representation was reduced from two to one, Newcastle, Berwick and Durham retained their two MPs, while Tynemouth, Sunderland, Gateshead and South Shields became new boroughs, each with a single MP. So, from having had eight urban MPs previously, now the North East had 12.

Popular support for reform had been embodied above all in Thomas Attwood's Birmingham Political Union, formed in 1830. This was soon followed in Newcastle by the foundation of the Northern Political Union. It was chaired by Attwood's brother Charles, a Gateshead glassmaker. Its members were left frustrated at the narrowness of the Act's changes, and working-class supporters felt betrayed by not having been given the vote. Grey had a fundamentally conservative approach to reform, seeing the Act as a long-term solution to parliamentary representation. No doubt many Newcastle voices must subsequently have expressed misgivings about the adulation paid to Grey and to the naming of 'his' street and the erection of 'his' monument paid for by public subscription. Perhaps there was some symbolism in the fact that a bolt of lightning in 1941 knocked Grey's head from the monument, with it having to be recreated from the fragments. Nevertheless, the 1832 Act was a singular achievement since forlorn attempts at reform had been made long before then. Its initial inroads into the unrepresentativeness of suffrage started an unstoppable process. Progressive extensions widened the franchise until universal suffrage was achieved, albeit well into the twentieth century.

Maps played a key role in determining many of the 1832 Act's decisions. The link with the military origins of detailed mapping was evident in the appointment of Lieutenant Robert Dawson, from the Ordnance Survey of Ireland, as supervisor of the Boundary Commission. This drew up new boundaries for all the boroughs and his maps traced out, where relevant, the boundaries of the old borough, the new borough and other administrative areas such as townships and parishes. Gateshead's map is typical. The parish was used as the basis of the new borough boundary, to which the small sliver of the Chapelry of Heworth was added. Gateshead at this date was

still a relatively small town of around 15,000 population, clinging to the south bank of the Tyne and with a straggle of development to the south along the line of the Great North Road, but small settlements such as High and Low Fell were beginning to expand and would eventually form significant parts of the town.

For Newcastle, existing townships were used as the basis for the extension of the borough's new boundaries, adding Elswick, Westgate, Jesmond, Heaton and Byker. The area of the town more than doubled from 2,700 to 5,739 acres. Dawson signed each of the maps, even though they were not drawn by him. He was an Ordnance Survey (OS) draughtsman and instructor for over 40 years. For the great majority of the Boundary Commission maps he was able to draw directly on existing published or draft OS maps. However, by 1832, OS coverage had not extended as far as the North East, so Dawson relied either on existing privately surveyed plans or on new surveys commissioned from private surveyors. A York-based surveyor, Robert Cooper, produced the 'Dawson' plans of South Shields and Tynemouth. The Newcastle and Gateshead plans were probably drawn from Thomas Oliver's maps since his huge 1830 plan of Newcastle is included in the Boundary Commission folders in the National Archives.

The debate over the boundary of the borough of South Shields was a lively one. A strongly argued case was made for the boundary to extend as far as the boundary with Gateshead, the commissioners suggesting that 'By the members of Gateshead and South Shields the immense interests of commerce fixed on the south bank of the Tyne would then be equally and similarly represented.' But as the map shows, a more restricted area was designated as the parliamentary borough, excluding Jarrow. This was largely due to the opposition of shipowners, anxious not to have their representation diluted by including a wider geographical area. The electorate

OPPOSITE: Robert K. Dawson, *South Shields: proposed borough boundary* (1832) [SANT] The first MP for South Shields, Robert Ingham, was initially elected as a Tory but was later returned as a Whig.

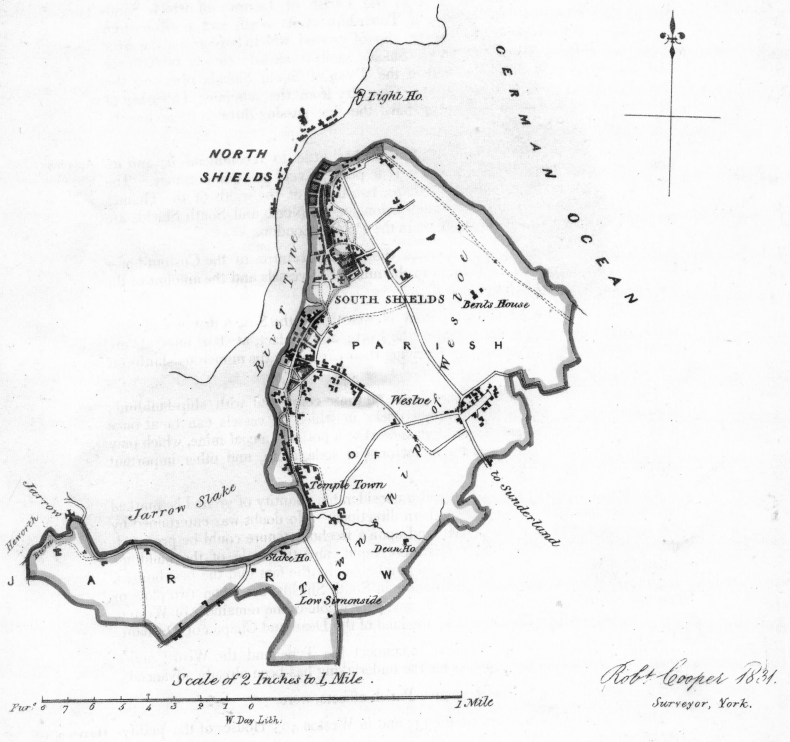

GERMAN OCEAN

Light Ho.

NORTH
SHIELDS

River Tyne

SOUTH SHIELDS

Bents House

P A R I S H

Westoe

W E S T O E

O F

Temple Town

to Sunderland

Jarrow Slake

Dean Ho.

Jarrow

Heworth

Burn

Stake Ho.

J A R R O W T O W N

Low Simonside

Scale of 2 Inches to 1 Mile.

Fur.s 8 7 6 5 4 3 2 1 0   1 Mile

W. Day Lith.

Robt Cooper 1831.

Surveyor, York.

Explanation.

Boundaries of Parishes or Townships Brown

Rivers _____ Blue

Proposed Boundary of Borough ___ Red

Robt. K Dawson

Lieut. R.E.

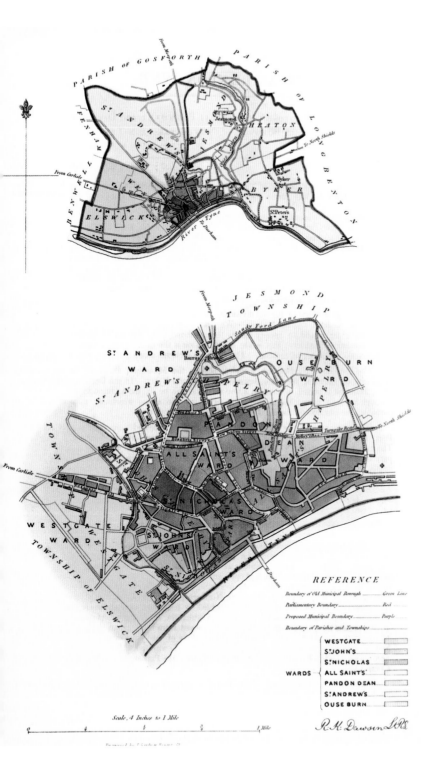

Scale, 4 Inches to 1 Mile

REFERENCE

Boundary of Old Municipal Borough .... Green Line
Parliamentary Boundary .... Red
Proposed Municipal Boundary .... Purple
Boundary of Parishes and Townships ....

WARDS
- WESTGATE
- ST JOHN'S
- ST NICHOLAS
- ALL SAINTS'
- PANDON DEAN
- ST ANDREW'S
- OUSE BURN

R.K. Dawson LRS

consisted of 495 individuals out of a population of around 19,000 (2.6 per cent). Robert Ingham of Westoe became the first elected MP.

Shortly after the Great Reform Act came the Municipal Corporations Act of 1835. Grey's government had set up a select committee in 1833 to inquire into abuses in many boroughs. Great variations were found in how Corporations were chosen and operated, but in over 180 boroughs that the committee members visited, Corporation members were the only people allowed to vote and they generally re-elected themselves or nominated friends and relatives, with the result that power was usually held by a small group of people.

Newcastle and Gateshead were amongst the boroughs which the committee visited, and they proved typical of the widespread malpractices. Newcastle had long been dominated by a small clique of influential individuals. The Hostmen comprised a cartel of burgesses who controlled coal exports from the Tyne. Almost all the governors of the Hostmen at one time or another became Mayor of Newcastle, MP for the town or Sheriff of Northumberland. The mayoralty was held repeatedly by a limited number of individuals: in the sixteenth century Robert Brandling served on five occasions and Henry Anderson on five; in the seventeenth century Sir Thomas Liddell served on four occasions; in the eighteenth century the Ridley family appeared seven times, the Blacketts no fewer than nine times and Nicholas Fenwick on five occasions. These were powerful individuals with interests in coal, shipping and glass, and with large land holdings. The Ridley family exerted an almost complete domination of the governance of the town. Sir Matthew Ridley, who had extensive interests in coal mining

LEFT: Robert K. Dawson, *Newcastle upon Tyne: parliamentary boundary and ward boundaries* (1837) [AUTH] Although the built-up area is accurately mapped and main streets named, the only individual building identified is the prison.

and in glass and soap manufacturing, was one of the town's two MPs from 1812 until his death in 1836. He succeeded his father in the seat which the family had held continuously since 1747. Nor was such domination restricted to the mayors and MPs. The role of town clerk was held by Nathaniel Clayton for 37 years from 1785 to 1822 when it was passed to his son John, who held the post until 1867 and played so vital a role in the reconstruction of the central part of the town. This was not an insignificant post since the clerk was automatically a member of the Corporation and all proposals for development in the town had to go through his office before going for approval to the Common Council.

The 1835 Act tackled such abuses by establishing a uniform system of municipal boroughs governed by town councils elected by ratepayers. Councillors were elected for three years, with one-third elected annually. Each borough was to have a paid town clerk and treasurer, neither of whom could be members of the Council. The election of councillors was clearly a major progressive change. In many of the larger boroughs the Commission recommended creating separate wards, and this was so for both Newcastle and Gateshead. Dawson's maps of the two towns depict their recommended ward boundaries. Gateshead was divided into three wards and Newcastle into seven. The size of each ward partly reflects their population density since the Commissioners aimed to create roughly equal numbers of ratepayers. In Gateshead, the East ward had 405 ratepayers, the West 397 and the South 480; while in Newcastle, Westgate had 648, St John's 499, St Nicholas 972, All Saints 790, Pandon Dene 1,028, St Andrew's 685 and Ouse Burn 643.

RIGHT: Robert K. Dawson, *Gateshead: proposed borough boundary and ward boundaries* (1837) [AUTH] Dawson did not personally survey every area, but his distinctive signature appears on all the maps produced for municipal reform.

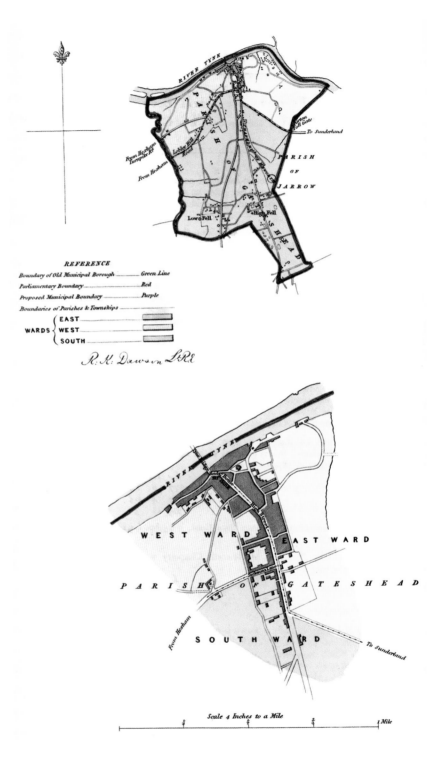

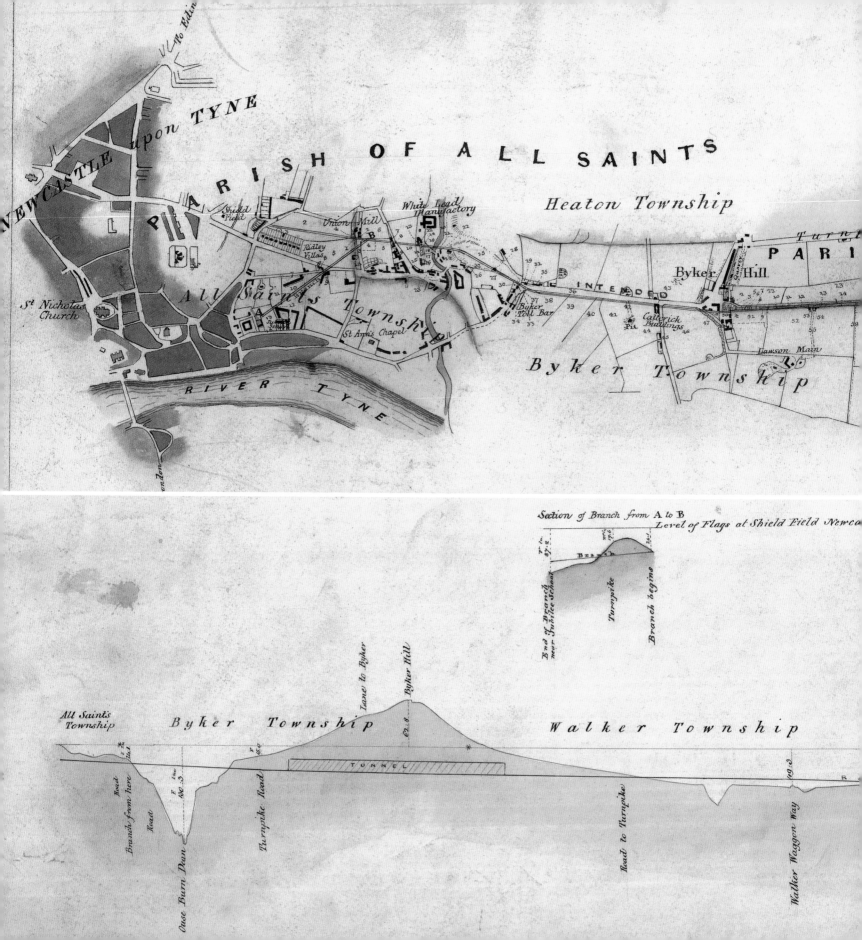

NEWCASTLE upon TYNE

PARISH OF ALL SAINTS

Heaton Township

White Lead Manufactory

Shield Field

Union Mill

Ridley Villas

All Saints Township

St Anns Chapel

St Nicholas Church

Byker Toll Bar

INTENDED

Byker Hill

Catterick Buildings

Pit

Lawson Main

Byker Township

PARI

Turn

RIVER TYNE

Section of Branch from A to B
Level of Flags at Shield Field Newca

End of Branch near Jubilee School

Branch

Turnpike

Branch begins

All Saint's Township

Byker Township

Lane to Byker

Byker Hill

Walker Township

TUNNEL

Branch from here

Ouse Burn Dean

Road

Turnpike Road

Road to Turnpike

Walker Waggon Way

1833

The Newcastle–North Shields railway

We indulge the hope that greater heights will be attained – and that posterity will look back to this age as one in which Railways and Locomotive Engines were only in their infancy!

J. Richardson, *Observations on the Proposed Railway from Newcastle upon Tyne to North Shields and Tynemouth*, 1831

Some local historians argue that the Newcastle–North Shields railway line was the world's first passenger railway. The case is a strong one, although goods traffic was also significant. In his *Observations . . .* of 1831 (basically a prospectus for the line), Joshua Richardson, one of the main protagonists, unequivocally stated '. . . on the Newcastle, North Shields and Tynemouth Railway, the principal revenue is to be derived from passengers'. This was an unusual expectation for any North

East railway proposal, but his data showed that the revenue from passengers would be eight times that from goods. The 1830s saw a frenzy of railway speculation in the Tyneside region, but many of the proposals concerned mineral lines and lines designed for mixed traffic. The exception was this proposal to link Newcastle to North Shields and Tynemouth. The idea was not new, but the earlier schemes were mainly speculative, and none had been properly surveyed. However, there was considerable disagreement over the precise route.

Robert Nicholson, a distinguished engineer who had surveyed the Liverpool–Manchester railway under George Stephenson and who was later to construct the Whittle Dean Water Works (*see* 1845), published the survey shown here in the lead map. His line started at Shieldfield, ran through Byker and Wallsend and terminated at Saville Street, North Shields. As this 1833 plan shows, Nicholson proposed a branch from

OPPOSITE: Robert Nicholson, *Plan and Section of a Proposed Railway from Newcastle upon Tyne to North Shields* (1833) [TWA]
The section shows the severe obstacle of the Ouseburn Valley and the proposed tunnel under Byker Hill.

just to the east of Ridley Villas towards the Quayside, terminating near the Keelmen's Hospital. Nicholson also proposed to dig a 1,600-yard tunnel under Byker Hill and a further branch from Percy Main to an intended dock at Coble Dean.

But diverging opinions on the course of the line remained and caused considerable delay with an influential group favouring a riverside route. Joshua Richardson, an articled pupil of Robert Stephenson, had surveyed the area in 1831 and the 1834 map shown here was, in effect, a revival of this scheme, starting at Stockbridge and proceeding through St Peters, St Anthonys and Carville. With the ultimate intention of linking to the Newcastle–Carlisle railway, a tunnel through Newcastle was also suggested, to link Stockbridge to Skinner Burn in the west. The promoters of these competing schemes appear to have come to an amicable agreement to subject them to independent assessment by Benjamin Thompson who had carried out the preliminary survey for the Newcastle–Carlisle line. Thompson reported in favour of Nicholson's inland line in 1834 although with a deviation to the north to avoid the tunnel under Byker Hill.

In his 1831 Report, Richardson had been at pains to set out an evidence-based demonstration of demand, with statistical data on traffic between North Shields and Newcastle. The flows were considerable: 36 steam packets, 25 gigs and 11 coaches travelled daily between the two towns in summer, carrying an estimated maximum of 896,000 passengers annually, while carts, keels and wherries carried an estimated 15,682 tons of goods. Nevertheless, the line was initially fiercely opposed by the river interests as a 'poverty producing measure' and by landowners fearing the loss of revenue from their wayleaves. Also, some North Shields businesses complained that the railway would induce shoppers to forsake local shops in favour of Newcastle, while Tynemouth lodging-house keepers feared that people would cease taking rooms in the village if they could take a train, sea-bathe and return to Newcastle within a few hours.

One key issue was whether the line should be solely for passengers or also include goods traffic. The Duke of Northumberland was particularly eager to exclude the carriage of coal

in order to protect his income from waggonways. On the other hand, commercial interests in North Shields objected that the exclusion of coal would deter the construction of docks. A compromise was reached whereby the railway company paid landowners ¾d per ton of coal per mile of the track through their land. This was significant in ensuring that the line finally adopted in 1835 was the more direct inland one through Heaton and Byker although lying a little to the north of Nicholson's 1833 proposal and terminating near Nile Street, North Shields rather than the original intention of Saville Street. The line required the construction of a spectacular viaduct across the Ouseburn Valley 918 feet long at a height of 108 feet, designed by John and Benjamin Green, as well as the 1,048 feet long and 82 feet high viaduct over Willington Dene. Both involved unique engineering techniques for the time and the whole line was acclaimed as one of the then foremost achievements in railway engineering.

The grand opening was on 18 June 1839, celebrated by two trains carrying 750 passengers, leaving from the temporary Carliol Square terminus (the terminus moved to Manors in 1842). The extension to Tynemouth was added in 1847 and by 1861 there were 25 trains each day operating between Newcastle and Tynemouth. A riverside branch, closely following Richardson's 1834 route through St Peter's, Walker and Willington Quay, was added in 1879 to serve mainly riverside industrial workers.

Inland from the river and coast the precursors of the Blyth and Tyne Company (formed 1853) were active, notably the Seghill railway. The main driver was to secure outlets on the Tyne for coal export, leading to the construction of a line from Blyth to Northumberland Dock. But this allowed a link to carry passengers on the Newcastle–North Shields line from the junction at Percy Main. In 1864 the Blyth and Tyne extended lines westwards from Backworth to Newcastle through Benton and South Gosforth and eastwards to Monkseaton (see 1841). Its Newcastle terminus was at Picton House, a villa built by John Dobson around 1830, with the station later becoming known as New Bridge Street. In its first week 17,000 passengers were carried, most of whom were in third class, suggesting

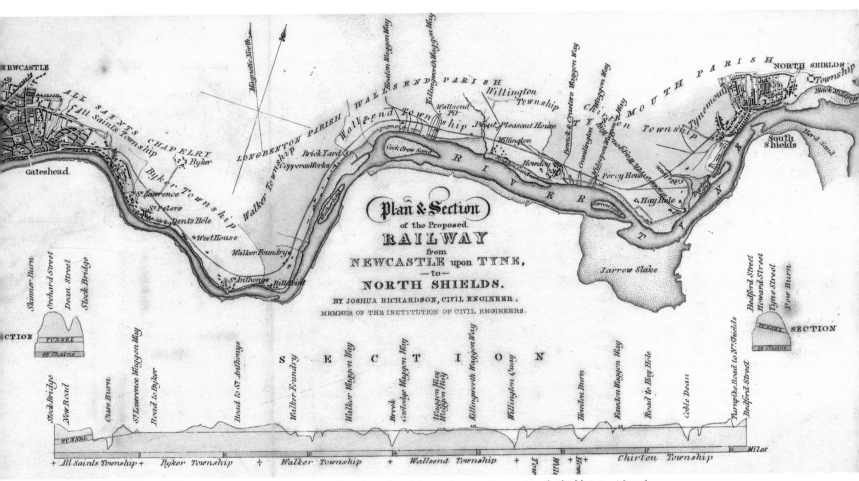

Joshua Richardson, *Plan and Section of the Proposed Railway from Newcastle upon Tyne to North Shields* (1834) [MIN]
The section shows the intended tunnel under Newcastle, to link with the Carlisle–Newcastle railway in the west.

the impact at this stage was mainly on ordinary working people rather than on middle-class commuters. The Blyth and Tyne railway competed effectively with the North Eastern Railway (NER) for traffic between Newcastle and North Shields and carried over a million passengers per year. It was taken over by the NER in 1874.

Thus, by the mid nineteenth century, with Manors station and the nearby New Bridge Street, Newcastle was the terminus for two important local routes north of the river, serving the eastern sector. The former linked Newcastle with riverside settlements, whilst the latter was instrumental in knitting together inland areas with the coast and Newcastle. By the 1860s the region had a maturing rail network. Several impressive new stations were built, a result of demand from two

relatively new sources – middle-class commuters and holiday makers. However, with the development of electric trams, short-distance suburban railways became relatively uncompetitive. For example, the number of passengers from Tynemouth in mid-May fell by 50 per cent between 1900 and 1902. The response was to electrify the suburban railway lines from Newcastle to the coast, a measure that proved highly successful and was an obvious precursor to the modern Metro system. One difference, however, was that the electrified passenger service shared the lines with substantial freight traffic, some 820,000 tons in 1920 for example. The Tyneside electric system carried nearly 13 million passengers in 1929 and, at a time when railway passengers were declining generally, succeeded in maintaining stability.

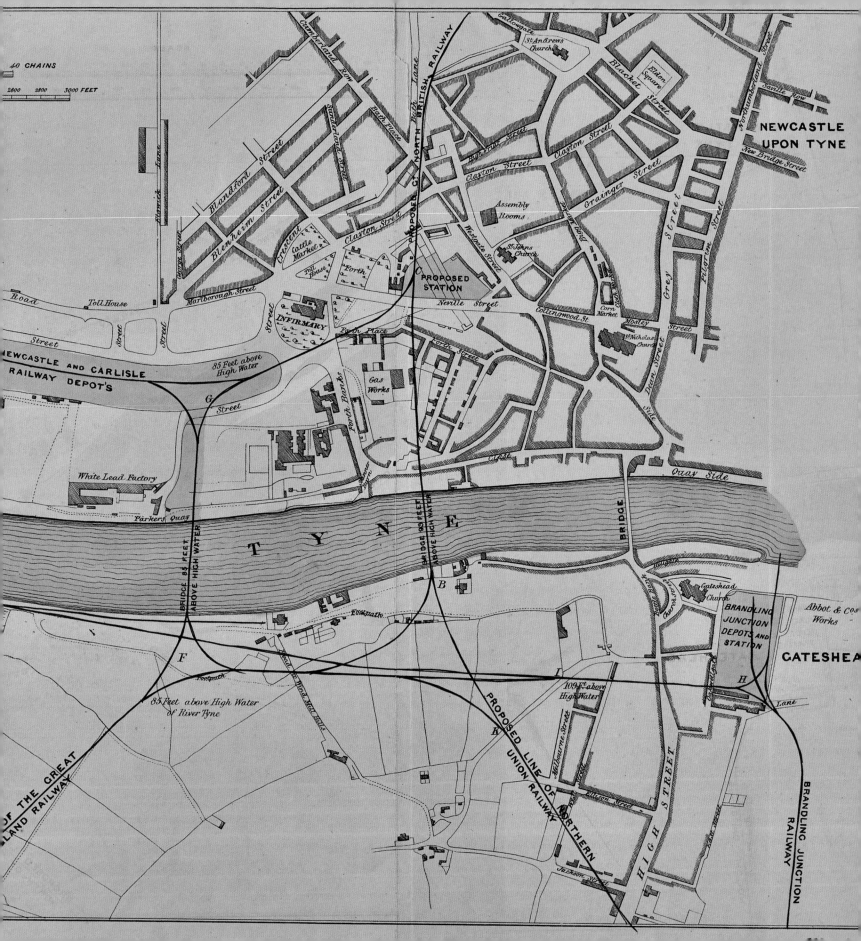

1841

The evolution of a rail network

The region's railway network . . . had begun life as a patchwork of small lines.

N. McCord, *North East England*, 1979

To modern eyes, Newcastle is the natural centre of the railway network of north-east England and the alignment of contemporary railway lines gives a superficial impression that the network was designed with this in mind. The evolution of the rail network around Newcastle, however, was much more complex and the result of many decisions and compromises made by individuals and small groups in relative isolation and responding to localised needs rather than a coherent infrastructural development. It was also a story of many false starts.

The point is well illustrated by the map opposite, representing one of at least ten attempts to cross the Tyne at Newcastle by railway. A first – and passing – glance suggests a familiar layout of railway lines between Newcastle and Gateshead, but a moment's reflection reveals that the lines on this map did not materialise as shown. These proposals came from two eminent engineers, Nicholas Wood (first president of the North of England Institute of Mechanical and Mining Engineers in 1852 until his death in 1865) and Thomas E. Harrison (later to be Robert Stephenson's assistant engineer in the building of the High Level Bridge). They involved building two railway bridges to create a union of two railway companies' lines in Newcastle at a station on the north side of Neville Street, and then linking this with the Newcastle–Carlisle route just to the west of the Infirmary and Forth Banks. The unified line (termed here the Great North British railway) would then proceed northwards up the west side of the city centre. But these plans were not executed, and it was to be at least a decade before something approaching a coherent network emerged.

OPPOSITE: Nicholas Wood and Thomas E. Harrison, *Plan of the different crossings of the River Tyne at Newcastle* (1841) [GLH]
With only one road bridge across the river at this time, the proposal to build two railway bridges was audacious to say the least.

One of the first lines to excite interest in longer-distance links was the Newcastle–Carlisle route. Although the Bill authorising the Carlisle line was passed in 1829, it took almost a decade before the line was fully opened due both to repeated changes of mind by the directors and to obstruction by landowners. Even by the time of its official opening on 18 June 1838, the Carlisle route's terminus was still on the Gateshead side of the river at Redheugh. Nevertheless, the opening was a memorable affair, with five trains arriving from Carlisle with its Mayor and Corporation who crossed the Tyne by boat and took breakfast with their Newcastle counterparts at the Assembly Rooms. The hospitality overran and, on returning to the trains with their number now much augmented, it was discovered that the Mayor and Corporation of Gateshead, plus a large and impatient crowd who had been waiting for over an hour, had commandeered the seats reserved for the Carlisle and Newcastle contingents. The waiting locomotives now numbered 14, pulling 130 coaches and carrying 3,500 people! Most of the coaches were uncovered and it began to rain heavily at Blaydon. The journey was accompanied by a

John R. Jobbins and Charles F. Cheffins, detail of *Brandlings Junction Railway* (c.1840s) [SANT] Despite the potentially lucrative mineral and passenger traffic, an inquiry in 1842 found that this railway was not paying its way.

thunderstorm and there was a collision between two of the engines, partially blocking the line. What was scheduled to be a four-hour journey took eight hours. Following the construction of the railway bridge over the Tyne at Scotswood, a temporary terminus was opened at Forth Banks in March 1847 and it was not until 1 January 1851 that the line was completed into the west end of the new Newcastle Central Station. By this time the High Level Bridge provided access to east coast trains on the London to Edinburgh route, which had to enter and leave the station at its eastern end.

Rather less fanfare had accompanied the development of lines south of the Tyne. The most significant of these was due to John and Robert William Brandling, substantial landowners who had pioneered deep coal mining. The Brandlings had leased substantial coal-producing areas in north Durham and were anxious to link these to the Tyne. They formed the Brandling Junction Railway in 1835, with the projected route passing through part of the Brandling's own Felling estate and then on to South Shields and Monkwearmouth, although with a complicated junction at Brockley Whins. The first section of the line to be opened in January 1839 was the incline from Redheugh, west of Gateshead centre, where the line was initially worked by a stationary engine. As the map shows, to the east of Gateshead the line picked up traffic from several

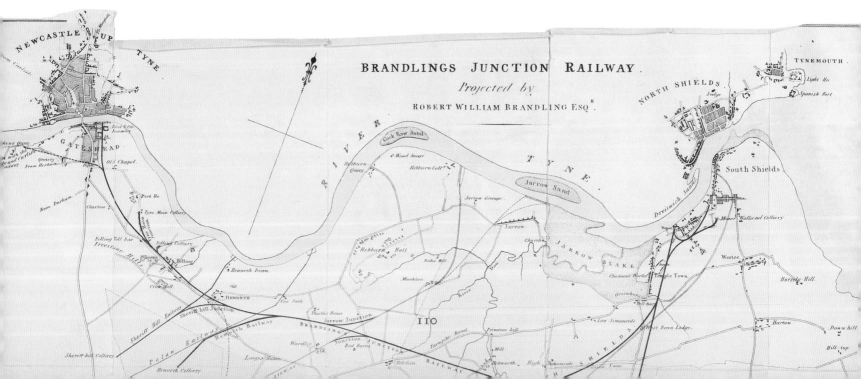

110

local railways en route. Although the origins of the line lay firmly with mineral extraction and conveyance, it proved to be surprisingly popular as a passenger route.

Reinforcing the point about the haphazard development of a regional rail network, the Blyth and Tyne was a classic case of a railway emerging from a large array of former horse-drawn waggonways around Seghill and Cramlington. The construction of a line in 1857 from Blyth to Northumberland Dock on the Tyne (*see* 1833) made possible a massive increase in the export of steam coal from south-east Northumberland. From this line, a short branch connected to Percy Main where it linked to the Newcastle–North Shields line. Additional branches were developed to shipping staithes along the river. A line from Hartley southwards to Tynemouth (with a different terminus just to the west of the Newcastle–North Shields line's station there) was opened in 1860, using part of the former Whitley waggonway and with a junction at Monkseaton (confusingly, then called 'Whitley') and a station at Cullercoats. This line ran to the west of the present line. In 1864 the Blyth and Tyne Company extended lines from Backworth west to Newcastle through Benton and South Gosforth, and east to Monkseaton – almost, but not quite, creating the now familiar circular route from Newcastle to the coast. It was not until 1882 that the full coastal loop was completed with the building of a magnificent new station at Tynemouth and the link up the coast through Cullercoats and Whitley Bay (where new stations were built further to the east). As the 1874 map shows, these three local developments, the Newcastle–North Shields line, the Brandling Junction line and the Blyth and Tyne, together with the Newcastle–Carlisle route, now provided a mature intra-regional network.

In terms of communications between the region and the rest of the country, Tyneside had for centuries enjoyed excellent maritime linkages, but the need for more efficient land links was obvious. Yet it was not until 1849 that a London–Edinburgh link was completed. Various routes were proposed and opposed; in Northumberland, Earl Grey didn't want it crossing his Howick estate, while the tradespeople of Morpeth wanted it to serve their town. A number of companies began

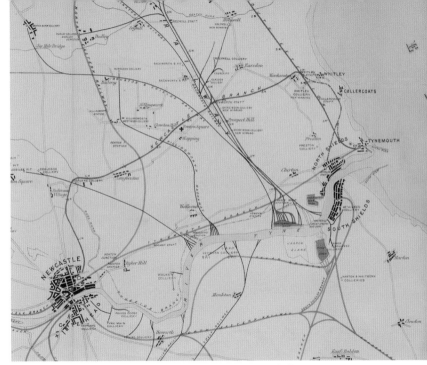

John F. Tone, detail from *Map of the North Eastern and Blyth & Tyne Railways* (1874) [SANT] Red = Blyth & Tyne lines; Blue = North Eastern Railway lines; Dotted Red = Blyth & Tyne lines authorised but not yet constructed; Black = other lines.

to work together to ensure a through line. By 1841, it was possible to reach Darlington from London via Leeds. By 1844 George Hudson had completed his Darlington to Gateshead line by purchasing the Brandling Junction line. The North British railway from Edinburgh to Berwick was completed by 1846. The missing link was from Berwick to Newcastle (or, more accurately, to Gateshead). The intense pressure to complete the east coast route pitted George Hudson (supported by George and Robert Stephenson) against Isambard Kingdom Brunel who proposed an atmospheric railway. Parliament favoured Hudson, and in 1845 construction began, with the line reaching Newcastle in 1847 and using the old Newcastle–North Shields terminal at Carliol Square. Stephenson's combined road and railway bridge, meeting the great physical challenge of crossing the Tyne, had already been decided on in 1844. Queen Victoria opened the High Level Bridge on 9 July 1849, completing the railway from north to south across the Tyne. This remained the only railway bridge across the Tyne at Newcastle until the much-needed King Edward VII Bridge was opened in 1906. By this time, over 800 trains were using the High Level Bridge each day.

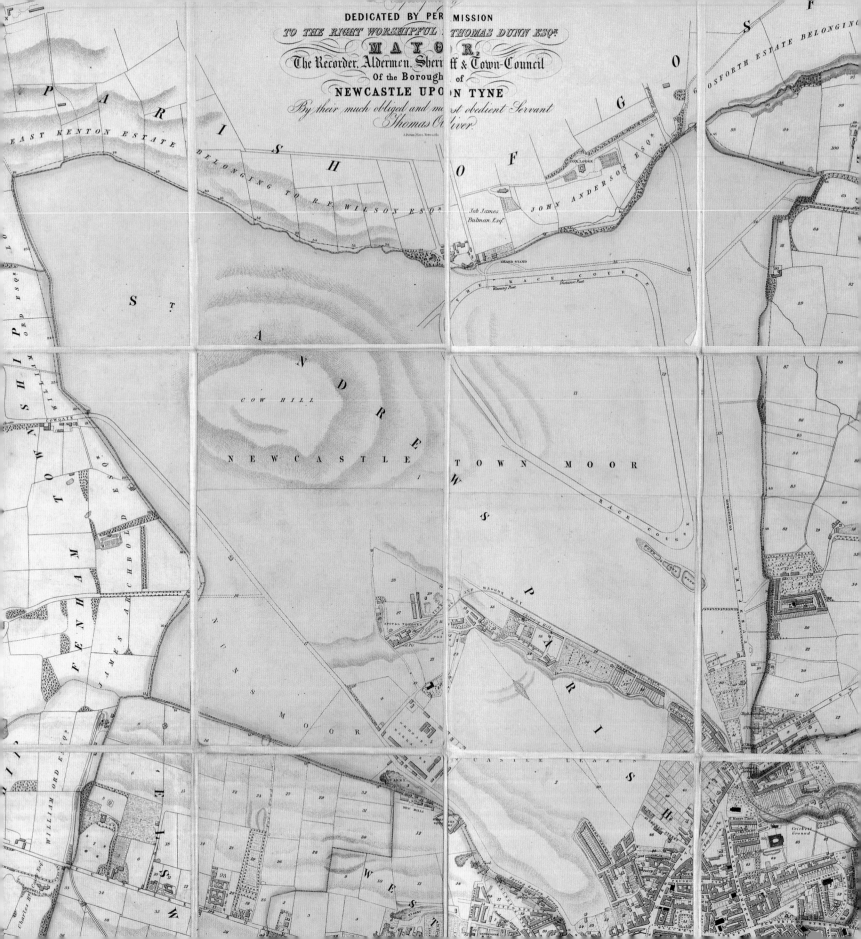

DEDICATED BY PERMISSION
TO THE RIGHT WORSHIPFUL THOMAS DUNN ESQ.
MAYOR,
The Recorder, Aldermen, Sheriff & Town-Council
Of the Borough of
NEWCASTLE UPON TYNE
By their much obliged and most obedient Servant
Thomas Oliver.

PARISH

EAST KENTON ESTATE

BELONGING TO R. WILSON ESQ.

GOSFORTH ESTATE BELONGING

COX LODGE

Job James
Bulman. Esq.

JOHN ANDERSON ESQ.

PARISH OF GO

ST. ANDREWS

COW HILL

NEWCASTLE TOWN MOOR

GRAND STAND

THE RACE COURSE

Winning Post   Distance Post

RACE COURSE

FENHAM

COWGATE

JAMES ARCHBOLD

WILLIAM ORD ESQ.

TOWNSHIP

NUNS MOOR

WAGGON WAY

SPITAL TONGUES

MOOR

NEWCASTLE LEAZES

WEST

WILLIAM ORD ESQ.

Charles Brandling Esq.

QUARRY

NEW MILLS

Crick Pit
Ground

1844

The continuous exploitation of the Town Moor

Fortunate among towns is Newcastle that she possesses, so near at hand, these open expanses of land.

R.J. Charleton, *The History of Newcastle upon Tyne*, 1885

Thomas Oliver's elegant map of 1844 shows the huge size of Newcastle's Town Moor in relation to its surrounding area. The Moor, in its entirety an area of almost a thousand acres, holds a very special place in the affections of Newcastle's citizens. It forms an integral component of the 'mental map' of Newcastle residents as an enormous green space only ten minutes' walk from the main shopping street. Words such as 'natural', 'unspoilt' and 'rural' are often used to depict the Town Moor, but such descriptions fail to recognise the complexity of its history and the processes that have produced the way it looks today. Strictly, it is not one 'Moor' but five –

Nuns Moor, Hunters Moor, Dukes Moor, Little Moor and the Town Moor proper. Furthermore, debate over its 'natural' state has been a recurrent feature over time. For example, no less a person than Lord Armstrong opposed a proposal for tree planting along the boundaries of the Moor in 1893 on the grounds that this would produce a tamed and artificial landscape and interfere with its natural state. But Thomas Oliver's 1844 map suggests that the Moor was hardly an untouched natural landscape even in the mid nineteenth century. For example, two major features of infrastructure are apparent in the form of the racecourse, complete with a grandstand, and the recently completed Victoria Tunnel and waggonway running from the coal pit at Spital Tongues to the quayside at the mouth of the Ouseburn. But even recognising these features underestimates the history and variety of exploitation of the Moor.

OPPOSITE: Thomas Oliver, detail of Town Moor from *Plan of the Borough of Newcastle* (1844) [TWA]
In 1833 the prestigious Northumberland Plate race was held for the first time.

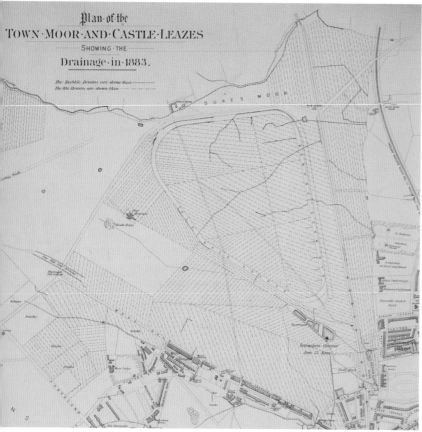

Property Office, Town Hall Newcastle, detail of *Plan of the Town Moor and Castle Leazes Drainage in 1883* (1883) [SANT]
For a 'moor' the investment in underground drainage is remarkable.

across the Moor there were several wayleaves related to coal extraction.

Even before the industrial period, coal was mined on the Moor, probably initially from simple drift mines but there are records of a pit being sunk in the mid sixteenth century, and in the early nineteenth century the Brandling family were awarded colliery rights there. During the Second World War, opencast coal mining took place on the Moor and, even during the 1970s construction of the Central Motorway running along its south-western edge (the North West Radial road), an opencast coal mining firm took advantage of the brief suspension of normal restrictions. The Town Moor is clearly anything but an 'unspoilt' natural landscape. Although superficially uniform, it also consists of a variety of different environments and features.

Following an attempt by Sir Walter Blackett MP in 1771 to enclose the Moor for the Corporation, the 1774 Act gave joint responsibility for the Moor to Newcastle Corporation and the Freemen of the town, with the former owning the soil and the hereditary Freemen having the right of herbage or pasture – a right that continues to be exercised today as the presence of grazing cattle in the summer months testifies. But its function as an unequivocal public space, albeit with some restrictions, was enshrined in an Act of 1925 which gave the public the right of access over its surface. Perhaps significantly, in recent years CCTV has been installed on parts of the Moor.

In reality, the Moor had functioned as a multi-purpose space for centuries. A charter granted by Henry VII in 1490 permitted an annual fair for the sale of livestock to be held. The site is shown on the 1883 map. Temporary army encampments were located there in periods of emergency: for example, General Wade camped on the Moor with 15,000 soldiers to protect Newcastle from the Jacobite army in 1745. It developed as a symbolic space for public fairs, and political and labour organisation demonstrations as well as public hangings were held there. In October 1831, a major demonstration of over 50,000 people in support of the reform movement caused considerable alarm, and on 28 June 1835 the army was called

Under a Parliamentary Act of 1774, the Corporation could lease up to 100 acres of the Town Moor for cultivation, although these leases were limited to seven years. Very few areas of the Moor were not affected by this process, although the horse-racing area became exempt from such leases. This agricultural exploitation led to improvement in the form of drainage, introduced as a distress-relief measure in 1842. This was subsequently extended as the 1883 'Drainage' map illustrates, indicating again the somewhat ambiguous nature of the Moor's 'natural' state. Its continuing role as a gathering ground for Newcastle's water supply is demonstrated by the 'ponds' (actually reservoirs, and later to become a boating lake), and in to counter a large Chartist gathering. Possibly the largest political gathering, however, was that which took place in favour of full male suffrage in 1873 when the crowd was said to exceed 200,000. Other forms of political activity are represented by the IRA's attempt in April 1921 to destroy Armstrong-Whitworth's airplane assembly works, which stood on the site of the old racecourse grandstand on the northern edge of the Moor.

But the use of the Moor for sporting events is probably one of its longest-standing roles. In the early nineteenth century, pitmen from south Northumberland frequently gathered for bowling matches on its eastern edge. A very different event took place on Easter Monday, 1822, when George Wilson claimed he could walk 90 miles within 24 hours. A section of the Moor was set aside, and 40,000 spectators were attracted. Watson achieved his objective, fuelled by mutton chops and warm gin, with 14 minutes to spare. Golf is another activity associated with the Moor and, while played informally for decades previously, was formalised in 1891 when a group petitioned the Freemen and Newcastle Council to allow them to create a proper course. The petition granted, the course was designed by the famous St Andrews golfer Tom Morris, and opened in 1892, with an unusual clubhouse established in one of the few remaining windmills in the city, Chimney Mills in Spital Tongues.

As the lead map suggests, probably the most significant sporting event was Race Week, which was held on the Moor from 1721 to 1881, having previously been held at Killingworth Moor and subsequently moving to Gosforth Park. Over its 160-year presence on the Town Moor, Race Week grew to be much more than a sporting occasion and, in fact, became the main public holiday for many Tyneside workers, with thousands journeying to the Moor to drink, gamble and enjoy the many sideshows and entertainments that accompanied the race meetings. Over 120 tents were erected to house these activities, and the main race, the Northumberland Plate, is recorded as attracting almost 50,000 people on occasions. Growing concern over public behaviour and drunkenness (the meeting was described by Charleton as 'the great saturnalia of the

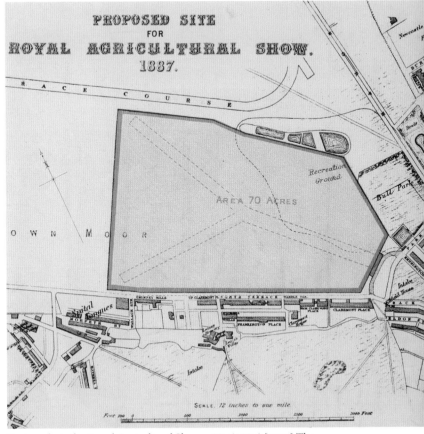

Proposed site for Royal Agricultural Show, 1887 (1887) [SANT] The extensive site of the 1887 RAS, Bull Park was the site of the Jubilee Exhibition in the same year.

North of England') lay behind the decision to move the meeting to Gosforth Park where it was much easier to impose formal controls and entry charges, while on the Town Moor the races were replaced by a Temperance Festival which became known as the Hoppings.

The Moor has also become the obvious site for large formal gatherings such as exhibitions. Agricultural shows, especially those entertaining national organisations such as the Royal Agricultural Society in 1887, became regular events. But rather more spectacular were the Exhibitions held to celebrate Queen Victoria's Golden Jubilee in 1887 and the North East Coast Exhibition of 1929 (*see* 1887).

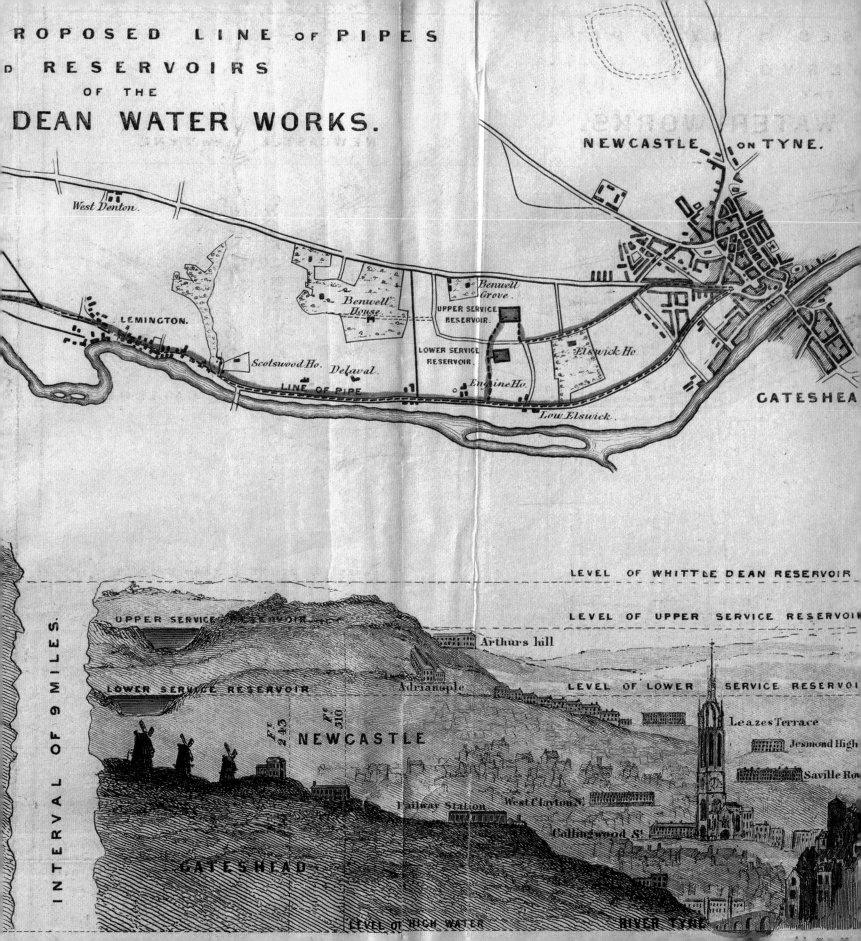

PROPOSED LINE of PIPES
D RESERVOIRS
OF THE
DEAN WATER WORKS.

NEWCASTLE on TYNE.

West Denton.

LEMINGTON.

Benwell House.

Benwell Grove

UPPER SERVICE RESERVOIR.

LOWER SERVICE RESERVOIR.

Scotswood Ho. Delaval.

Elswick Ho.

GATESHEA

LINE OF PIPE

Engine Ho.

Low Elswick.

INTERVAL OF 9 MILES.

LEVEL OF WHITTLE DEAN RESERVOIR

LEVEL OF UPPER SERVICE RESERVOIR

UPPER SERVICE RESERVOIR

Arthurs hill

LOWER SERVICE RESERVOIR

Adrianople

LEVEL OF LOWER SERVICE RESERVOIR

Leazes Terrace

Jesmond High

Ft 243

Ft 310

NEWCASTLE

Saville Row

Railway Station

West Clayton St.

Collingwood St.

GATESHEAD

LEVEL OF HIGH WATER

RIVER TYNE

1845

Supplying water and developing a sewerage system

. . . the water bottles on the table were filled with the Company's water, sparkling as from the fountain and clear as distilled by the alembic of nature.

Newcastle Journal, 28 October 1848

So claimed this newspaper when reporting on the dinner held to celebrate the inauguration of the Whittle Dean water supply system shown on the map. This was eventually to facilitate a sufficient clean water supply to the town, but only after much subsequent struggle and debate. In 1845, Dr D.B. Reid's report on the *Health of Towns* suggested that there were 24 'pants' (street fountains) in Newcastle, but only one in eight houses were directly supplied with water. The other Tyneside towns fared no better. In 1843 only 110 houses in Gateshead had water laid on, despite its population of 38,747: six standpipes

were the main supply there, supplemented by water sold from barrels carried around by carts. In South Shields in 1845, of the 3,911 inhabited houses, only 180 had water laid on, and 977 obtained it from the pants by paying a weekly charge to the water company. The majority relied on water carts. But in the same year as Dr Reid's report, the Whittle Dean Water Company was founded to supply Newcastle, but with only a partial commitment to meet domestic demand. As the map shows, the company intended to expand supply with the construction of a large impounding reservoir at Whittle Dean, 12 miles to the north-west (later increased to five reservoirs), along with two small service reservoirs at Elswick. This added substantially to the town's existing water supply taken from the Tyne at Ryton, from springs at Dunstonhaugh and from the River Pont at Dissington. Nevertheless, the supply fell far

---

OPPOSITE: Thomas Sopwith and Scott, detail from *Plan of the proposed line of pipes and reservoirs of the Whittle Dean Water Works* (1845) [ROB] The service reservoirs at Elswick were designed to pump water to the higher districts of Newcastle.

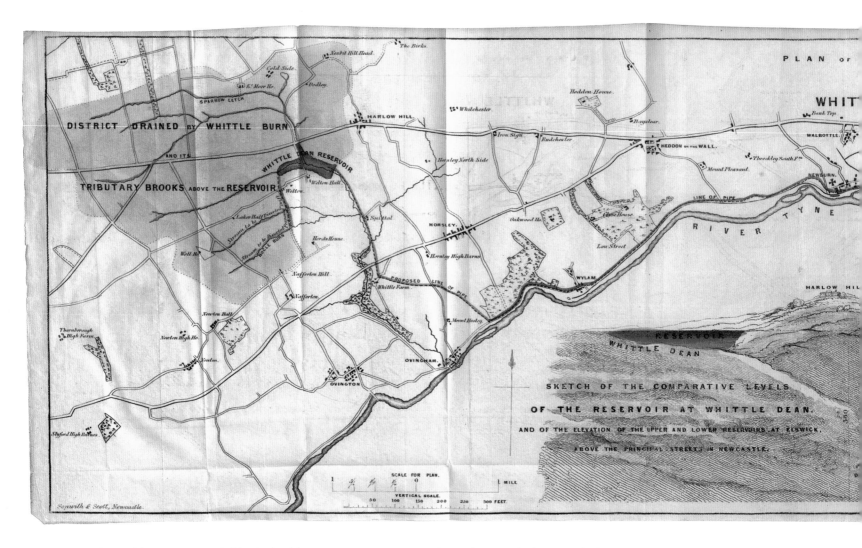

short of that recommended by the health commissioners. Water was still not seen as a public good.

In fact, meeting domestic consumption was the lowest priority for commercial water companies: most were established to supply industrial concerns, especially breweries. In North Shields in 1846 there were 26 breweries or other manufacturers with a direct water supply, but only 406 of the 3,225 houses were supplied directly. In Newcastle, the first commercial provider of water was the Newcastle Fire Office in 1797, a fire insurance company mainly concerned with providing water to extinguish fires. It did create a new source

by sinking a shaft at Coxlodge and accessing flooded coal workings, enabling it to supply water for two days a week. To meet the demands of a growing population, it took water from the Tyne and sold it from handcarts. This coincided with the arrival of the first cholera epidemic in the town in 1831, although the connection was not recognised at the time. Only gradually did the provision of clean drinking water and the means of disposing of human and other waste come to be seen in a different light, largely because of disease and epidemics. Cholera was no respecter of social class, adding political impetus to questions of sanitary improvement. Thus, although

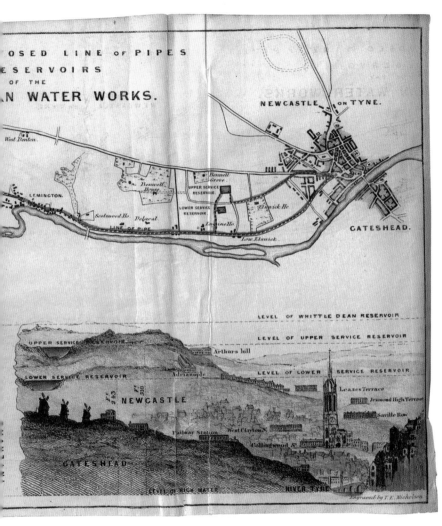

Sopwith and Scott (1845) [ROB] James Simpson, engineer to the Whittle Dean Water Company, exaggeratedly described the pipeline as 'the longest aqueduct ever known'.

discoverer of the source of cholera in London who had trained in Newcastle) noted that during the 1849 epidemic, Newcastle and Gateshead had 'escaped with a very slight visitation from cholera', whereas in North Shields, still largely dependent on the Tyne for water, the number of deaths was high.

But the company seriously underestimated demand. In 1853 it was only supplying 15,000 tenants out of a total of more than 115,000. It was decided to expand the Whittle Dean complex by building the Great Northern Reservoir. Demand continued to expand rapidly, and the company resumed pumping from the Tyne at Elswick. Soon after, disaster struck with the 1853 cholera epidemic, much more severe than that of 1849. The company experienced significant public criticism. Its response was to expand its catchment area, and over the next half century it built a chain of reservoirs to the north-west of Newcastle, connected by tunnels and aqueducts. In 1858, 23,000 tenants were supplied, and by 1881 the total population served by what had, from 1863, become the Newcastle and Gateshead Water Co., was 369,940. Importantly, the supply was of a better quality thanks to developments in the 1870s, such as the filtration plant at Throckley. However, a survey of 1883–85 found that a piped indoor supply was available to only 33.1 per cent of households in Newcastle and outdoor water supply to another 32.7 per cent, with only 6.7 per cent of tenements having an indoor tap and 23.7 per cent access to an outdoor tap. By 1897 more than two-thirds of England's local authorities had municipalised their water supply, but Newcastle was not among them.

In the mid nineteenth century it was still considered acceptable for effluent of all kinds to be disposed of via streams and drainage channels flowing to the Tyne. Although Tyne water was pumped out upstream at Elswick, that source was near the town and, more significantly, since the river was tidal, polluted water could be washed upstream. The sewerage map

the industrial demand for water was still the main consideration for the Whittle Dean Water Company, public health concerns were growing. Directors of the company included four former mayors of Newcastle and Sir William Armstrong. Their interests were largely directed at increasing dividends and fees rather than improving quality and quantity. Nevertheless, the company's AGM of 1847 reported that an additional 1,150 households had been supplied and one year later the first of the Whittle Dean reservoirs was opened. This allowed the company, at last, to cease pumping water from the Tyne at Elswick. No less an authority than Dr John Snow (the

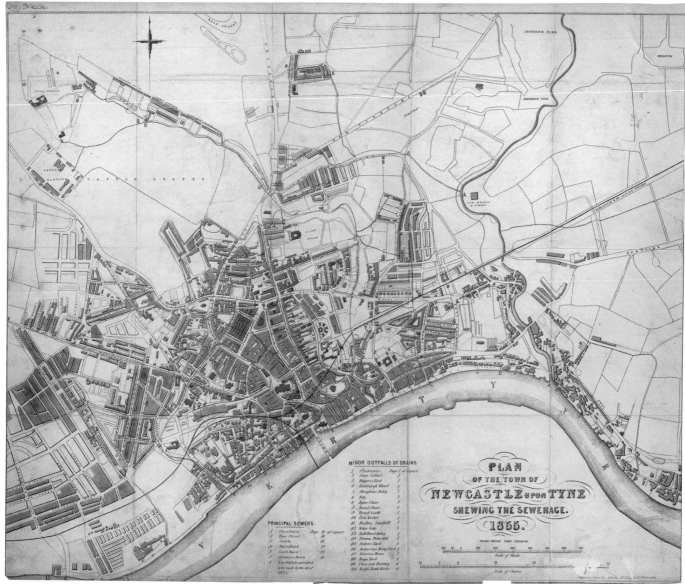

T. Bryson, *Plan of the town of Newcastle upon Tyne shewing sewerage, 1855* (1855) [NCL]

of 1855 illustrates the system which consisted of six 'principal' sewers, all of them ancient streams flowing into the Tyne, from Skinner Burn down to Ouseburn. These were augmented by 20 minor outfalls that also drained into the river. A campaigning medical officer, Dr Robinson, noted in 1857 that the sewerage system was a patchwork, with some larger sewers emptying into smaller ones and others laid at inappropriate levels and simply discharging into the ground.

Meanwhile, fearing the expense of creating a comprehensive system, the Council asserted that the town was adequately sewered. Yet, as the detailed map shows, the Sandgate area, while housing 5,000 people in just 350 dwellings had only two 'minor' drains with the partially culverted stream of the Swirle as its main outlet, and a sewer running along the Sandgate itself. Ironically, this was the location of a structure called 'The Folly' where Cuthbert Dykes had attempted to develop a water works to supply the town first in 1693 then again in 1736. Even more significantly, the whole area had only four private WCs, and one public and three private privies. The only water supply came from two pants. In contrast, in the newly built 'Grainger Town', 414 households with a total population of 2,070 had the use of 517 WCs.

As we have seen, Newcastle was one of the places most affected by the 1853 cholera epidemic and, although there was some immediate reaction such as the appointment of a superintendent of scavenging, there was little real progress in sanitation. National Improvement Acts in 1846 and 1850 had contained clauses compelling the provision of privies where none existed and requiring new-build houses to have privies and tenement properties to have proper drainage and sewerage. Unfit dwellings could be condemned and closed. But these provisions were rarely applied. The Health of Towns Commission of 1845 had found that new streets in Newcastle and Gateshead were as bad as the older ones. For example, G.T. Gibson, a solicitor specialising in property development and management, developed 'Gibson Town' on the eastern side of the centre in the 1840s and early 1850s (see 1830), but several of his streets still lacked drainage even by the 1860s. The area was drained principally by the Swirle. This inadequate sewer was fed by small sewerage pipes from Gibson Street, Blagdon Street and Howard Street, even before it reached the Sandgate area.

National legislation forced a reluctant Council to take responsibility for improvements such as appointing a Medical Officer of Health. The provision of conveniences improved between 1866 and 1885, with the gradual elimination of houses without any form of convenience. But problems persisted as toilet facilities were usually either a privy or a dry earth/ash closet, usually outside and shared. Street cleaning became an important urban function. While in 1833 the job had been contracted out for five years to six men with three horses and

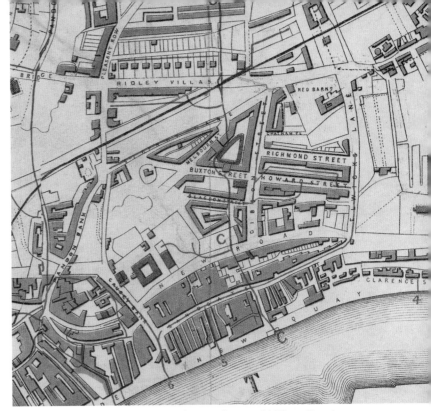

T. Bryson (1855) Detail of sewerage plan: Sandgate and 'Gibson Town'.

carts, and a further eight men sweeping, watering and collecting refuse, by 1907 the job was being done by the Engineer's Department with a total of 540 employees and 93 horses.

Despite progress in the latter part of the nineteenth century, problems remained. The length of sewer pipes increased from 78 to 227 miles between 1889 and 1913, but of the 4,023 new conveniences built between 1886 and 1888, while 78 per cent had water closets, the vast majority were still external. In 1898, Henry Armstrong asked the Sanitary Committee to insist that privies should be replaced by water or pail closets. The Sanitary Committee chose pail closets which cost £2 compared to £4 for a WC and were exempt from water rates. A survey of 55,000 houses in 1888 by the *Daily Leader* found that one-eighth were without water and one-fifth without any form of privy. This legacy of the town's overcrowded and inadequately serviced dwellings was to prompt one of the most ambitious public housebuilding programmes of the twentieth century. Meanwhile, the provision of public conveniences caused great debate in the Council, including some outrage at the cost (£94 'fixed up complete') of what, admittedly, was a rather superior gentlemen's urinal.

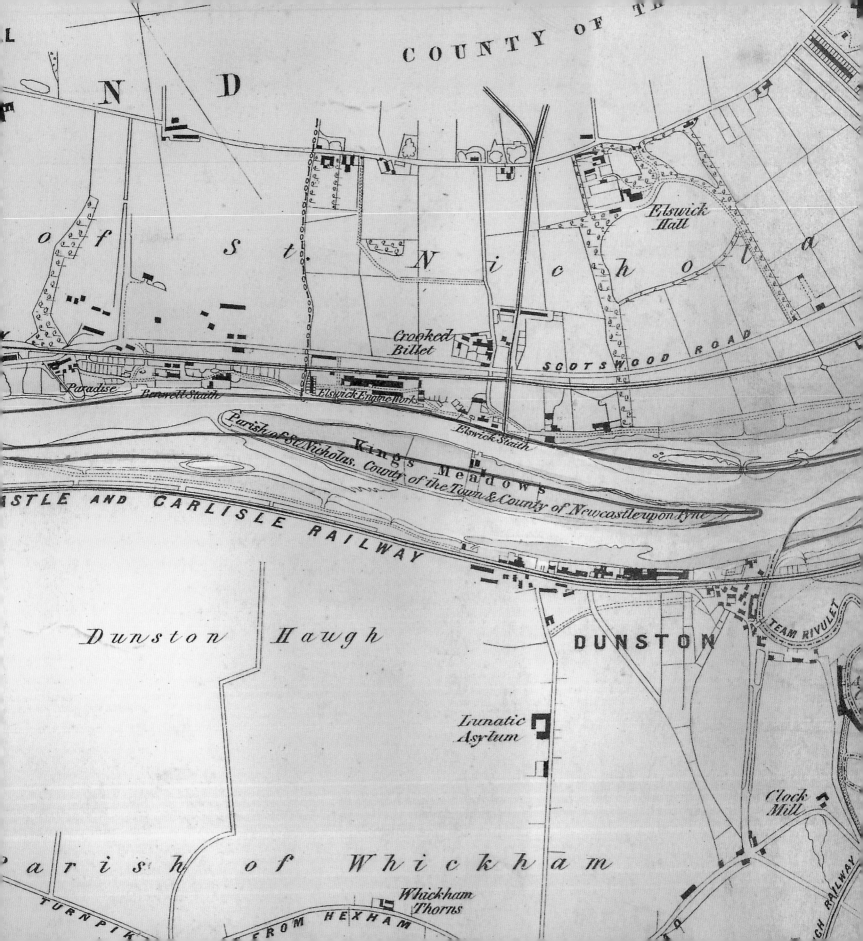

c.1852

W.G. Armstrong

A unique combination of scientist, engineer and businessman . . . a keen student of natural history, an accomplished landscape gardener and a generous philanthropist.

P. McKenzie, *W.G. Armstrong*, 1983

Sir William Armstrong is by far the most frequently referenced figure in the context of nineteenth-century Tyneside, but his achievements far exceed a mere numerical assessment. Trained as a solicitor, he became an inventor, a practical engineer and a businessman, and excelled in each. His firm was one of the largest engineering and armaments enterprises in the world. His life spanned virtually the whole of the nineteenth century. Born in 1810 in Newcastle's Shieldfield, he died in 1900, just a few months before Queen Victoria. His father was a corn merchant, active and well-connected in Newcastle's commercial circles and a local politician elected Mayor of Newcastle in 1850. These local contacts were vital in Armstrong's first commercial venture, the establishment of the Whittle Dene Water Company, of which he was secretary. Established to provide at least part of Newcastle with a continuous supply of mains water at a good pressure, Armstrong made no secret of his additional motive which was to investigate various commercial applications of hydraulic power.

J.T.W. Bell's splendid *Plan of the River Tyne* is undated, but it shows the suggested river 'improvements' of Brooks and Rendel, published in late 1851. There is no sign of Northumberland Dock, opened in 1857 and taking four years to build, suggesting the map dates to 1852 or 1853. Armstrong's 'Elswick Engine Works', built after his purchase of seven acres of land from Richard Grainger in 1847, is shown on the detailed map. Armstrong, with his partners, established a small

OPPOSITE: J.T.W. Bell, 'Elswick Engine Works' from *Plan of the River Tyne* (1852–53) [N.EST]
Note Kings Meadows and the general absence of housing. The red and black lines show proposals for the future channels of the river.

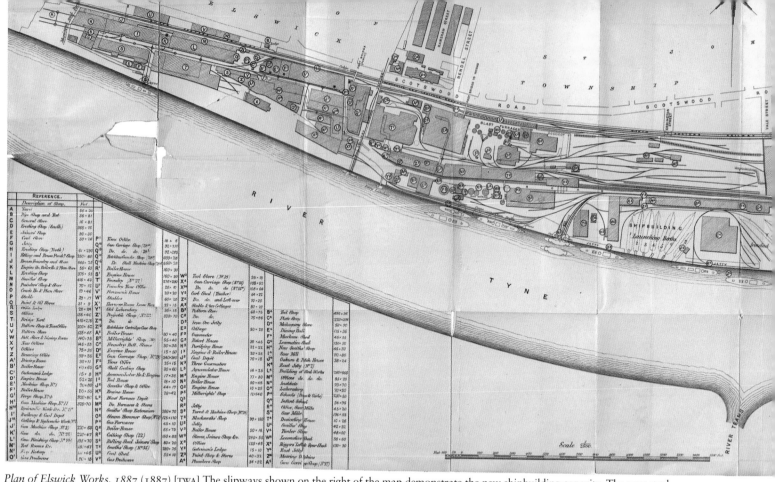

Plan of Elswick Works, 1887 (1887) [TWA] The slipways shown on the right of the map demonstrate the new shipbuilding capacity. The new steel works is just visible on the extreme right.

riverside workshop manufacturing a product he had invented – hydraulic cranes. In fact, the Elswick works had been longer in gestation than is usually recognised: in 1837, while still practising as a solicitor, Armstrong commissioned Thomas Sopwith to survey part of Grainger's estate with a view to establishing a factory there. Initially employing 20 or 30 men, by 1851 the workforce was just over 400, and by 1863 just under 4,000. By the end of the century Armstrong's works, now diversified into various branches of heavy engineering and shipbuilding, extended for three miles along the north bank of the river from Elswick to Scotswood and employed over 13,000 men. Such growth must have been inconceivable at mid-century, even to Armstrong himself. As Bell's map shows, the newly formed Tyne Improvement Commission (TIC) had invited proposals for improving the Tyne as a commercial river. At this stage, this stretch of the river was hardly a propitious location for large-scale industrialisation. J.M. Rendel (later to

be a significant business associate of Armstrong) proposed to channel the river on the northern side (shown by the black lines) and remove a substantial part of Kings Meadows. W.A. Brooks was less interventionist, proposing that the channel be deepened to the east of Kings Meadows (shown by the red lines) but leaving the island intact. There was no immediate action, but the nature of the immediate environment in 1852–53 suggests an unlikely site for a global industrial powerhouse.

The initial works consisted of a machine shop, boiler shop, smiths' shop and an erecting shop with joiners and pattern makers. A brief venture into the steam locomotive business was not successful, and the firm concentrated on hydraulic machinery. The first orders came from Liverpool's Albert Docks, and 152 cranes were manufactured between 1848 and 1852, with the first export order coming from Rio de Janeiro. Prophetically for a future local development, enquiries were received concerning the manufacture of a hydraulic-driven swing bridge for Birkenhead Docks.

A major addition to Elswick's manufacturing capacity came with the Crimean War and the need for modern field artillery. Armstrong developed a new rifled breech-loading gun, superior to current ordnance in both distance and accuracy. A grateful government offered him a knighthood in 1859 and the post of engineer of the Royal Gun Factory at Woolwich. The capability of the Elswick works in the armaments field was enhanced when Captain Andrew Noble, an expert on explosives and ballistics, joined the firm in 1861. Armstrong was a central figure in organising the British Association's visit to the North East in 1863, orchestrating lectures and demonstrations of the application of science to manufacturing. He edited the book on the region's 'Industrial Resources' which, for decades, provided the template for the annual publication celebrating the British Association's visits to towns and cities across the country.

In 1867, Armstrong's firm significantly diversified, working in association with the shipbuilding firm of Charles Mitchell at Low Walker. The idea was that Mitchell's would build ships and Armstrong's would provide the guns. However, with Armstrong himself becoming more interested in warship construction, the two firms merged in 1882, and shipbuilding was developed at Elswick, the Walker yard concentrating on commercial vessels and Elswick on warships. Shipbuilding at Elswick was made possible by major dredging works that deepened the Tyne and by the construction of the Swing Bridge that allowed access for large vessels upstream beyond Newcastle. The design of the latter is often mistakenly wholly attributed to Armstrong, but he worked alongside John Ure, the Tyne Improvement Commission's engineer. However, the Elswick works manufactured the structure. Between 1885 and 1903, 92 warships were built at the Elswick yard amounting to 234,661 tons. The intense development of the site even before this is shown in the 1887 plan drawn at the time of the Jubilee Exhibition. Over 180 named functional units were packed onto the site. Shipbuilding continued at Elswick until 1917 when the larger warships now demanded by navies could no longer squeeze past the Swing Bridge. What had been a solution in a previous age had now become a problem. The

only international rival to the Elswick works in the armaments trade was Krupps, which was heavily subsidised by the German government. Armstrong's, although benefitting from government contracts, always remained independent. The firm also continued to supply machinery for a more peaceful market, most notably providing the hydraulic pumping engines that operated London's Tower Bridge.

Armstrong's amazing progression was not entirely without adversity. Despite the superiority of his weaponry, the extreme conservatism of senior officers in the British army led to the loss of the War Department contract to supply ordnance in 1862. In the Tyneside Engineer's Strike of 1871 (over the refusal of the employers to consider the working day be cut to nine hours), he led the employers' side, but lost much local support through his stubbornness and also his enthusiasm for recruiting foreign replacement labour. Possibly not unrelated, he failed dismally (as a Liberal Unionist candidate for Newcastle) in an ill-advised venture into national politics in the 1886 election. In 1896, he caused a major rift in the Lit & Phil and resigned as president through a rather naïve (although typical of the age) assertion that politics should not come into debates about economic matters.

In 1897 Armstrong's merged with his erstwhile rival Joseph Whitworth and in the twentieth century Armstrong-Whitworth began to manufacture aircraft in Gosforth, on the edge of the Town Moor, continuing the success of Armstrong's manufacturing prowess. He was a remarkable inventor and engineer. The output of hydraulic machinery, armaments and shipbuilding from his huge Elswick works made Armstrong a rich man. Some of this wealth he used to benefit Tyneside with charitable giving, for example in donating his estate in Jesmond Dene to create Armstrong Park (*see* p. 140), founding Armstrong College (the basis of Newcastle University) and endowing the Hancock Museum of Natural History and local hospitals, including the Royal Victoria Infirmary. When he moved to live at Cragside in rural Northumberland, his use of hydraulics made that the first house in the world to be lit by hydroelectric power. He and his Elswick works were one of the glories of Tyneside.

1854

The Great Fire

There being none to parallel the late calamity in the amount of its casualties, nor in the extent of property destroyed . . .

'J.R.', *A Record of the Great Fire in Newcastle and Gateshead*, 1855

Fires were a frequent and much feared occurrence in big towns. The combination of coal-fired power generation, the use of chemicals, the legacy of wooden buildings from an earlier age and the inadequacy of fire-fighting capacity proved a lethal mix. Most fires were restricted to one or, at most, a few neighbouring buildings, but more widespread damage was not infrequent. Newcastle had had its share of serious fires before the nineteenth century: in 1284 the wooden medieval bridge was destroyed by fire; in 1644 local inhabitants set fire to buildings in Sandgate to prevent General Leslie from leading his Scottish army into the town; in 1725 a fire in the Side caused a barrel of gunpowder to explode, killing a dozen people; in 1750 the heat from a furnace in a brewery in the Close started a fire in the adjacent warehouse which led to the destruction of a number of houses and other warehouses. Overall, between 1720 and 1870 the town saw 168 serious fires, defined as those that resulted in either total destruction of property or loss of life. The majority of these took place in factories, workshops or warehouses. A number of these were attributed to arson and to bogus insurance claims.

However, the largest conflagration – the Great Fire – occurred in 1854. As Oliver's map shows, it caused devastation in both Newcastle and Gateshead with extensive loss of life and property. An 1855 newspaper report gives a graphic blow-by-blow account of the tragedy. The fire started shortly after midnight on 6 October in a worsted mill in Hillgate in

OPPOSITE: Thomas Oliver, *Plan of part of Newcastle and Gateshead shewing the property burnt by the Great Fire* (1854) [TWM] Buildings shown in red were destroyed but many others were damaged.

Bertram's Warehouse, Worsted factory etc (1855) [ROB] By 3 a.m. the warehouse had become a 'cataract of fire'.

Gateshead (number 8 property on Oliver's map). On Hillgate's southern side was a dense mass of poor tenements and houses, while on its northern side were numerous warehouses and industrial buildings with frontages abutting on the river. The fire spread rapidly through the mill, accelerated by the large quantities of oil used in the manufacturing process, and the building was soon reduced to rubble. Close by was a large six-storey warehouse – Bertram's Warehouse, named after its owner Charles Bertram (number 9 property on the map) – in which was stored a variety of chemicals and other combustibles. The chemicals included large quantities of sulphur, and the intense heat from the mill fire caused the sulphur to melt, catch fire and stream from the upper floors of the warehouse. Within a few hours a series of other warehouses had caught fire and the best efforts of the firemen to quench the flames proved of little avail. At about 3 a.m. there was a huge explosion which sent burning debris, rocks, stones and other projectiles hurtling into the sky destroying neighbouring warehouses and blocks of houses. The explosion was reported to have been heard 30 to 40 miles away in Alnwick to the north and Hartlepool to the south. Much of the east side of the lower part of Gateshead was demolished. Gateshead's parish church, St Mary's, was also wrecked.

Newcastle also suffered as burning embers from the explo-

sion were thrown across the river. Shops, offices and warehouses fronting much of the Quayside were set ablaze, again accelerated by the combustible contents of some of its warehouses, including paper and furniture. A third fire broke out in the dense tenement housing between Pilgrim Street and Butcher Bank. Most of the shops in Sandhill had their fronts blown out, and there was damage to properties in the Side, Pilgrim Street, Market Street, Dean Street, Collingwood Street and Mosley Street. Since the fire had started across the river in Gateshead, most of the few available fire engines from Newcastle had gone there initially and were unable to return across the river to tackle the fires in Newcastle. Later, in the course of the night and morning, more fire crews and engines arrived from as far afield as Hexham, Morpeth and even Carlisle and Berwick. Floating fire engines from South Shields and Sunderland were brought up the river to augment the water pumped onto the blazes.

The panorama gives a graphic impression of the scale and drama of the conflagration. Inevitably it had attracted large numbers of spectators and many – on the Quayside in Newcastle and on Hillgate in Gateshead – were injured by the explosion. Overall, it was estimated that between 400 and 500 people were seriously injured, including many firemen and soldiers who had been drafted in to tackle the fires and resulting chaos. Almost 50 people are thought to have been killed, mainly occupants of the poorer houses and tenements.

But it was not just the poor who suffered. Also among the dead was Charles Bertram in whose bonded warehouse the massive explosion had occurred and who was also a magistrate. Another was Robert Pattinson, one of the sons of the owner of the tannery of Pattinson & Sons and himself a councillor for Westgate Ward. He died in bizarre circumstances since the mayor of Gateshead had been in Hillgate but, wanting also to inspect Church Walk, asked Pattinson to stand in for him in Hillgate just before the explosion. The mayor survived; Pattinson did not.

A further poignant death was that of Alexander Dobson, the second son of the Newcastle architect John Dobson. Alexander was 26 and at the start of what would undoubtedly

M. and M.W. Lambert, *View of fire from the High Level Bridge* (1854) [ROB] The drawing is from the Lamberts' book on the Great Fire. They donated all proceeds to the families of the victims.

have been a distinguished architectural career. He had initially worked in his father's office, then moved to London to work with other practitioners and returned to Newcastle in 1852 where he again worked in his father's practice. His death may have been one of the reasons why the Dobson and Oliver families were at odds, since the warehouse that exploded and caused Alexander's death had been designed by Thomas Oliver.

Both Oliver and Dobson submitted schemes for reconstruction, possibly further evidence of their estrangement. Although their proposals for the immediate Quayside area were broadly similar, Dobson proposed much clearer linkages between the Quayside and the upper town centre, through a new street to meet Pilgrim Street and an extension of the Side to the northwest. Oliver, on the other hand appears to have had a greater focus on the East Quayside and linkages into the recently built 'Gibson Town', of which, of course, he was the architect. Dobson's plan created the impressive commercial buildings in Queen Street and Lombard Street, but his foresight in attempting to link the 'new', post-fire Quayside with the upper town with new access roads was ignored.

The inadequacy of largely voluntary arrangements for tackling fires was fully exposed by the 1854 conflagration. Around that time there were no more than a dozen fire engines in Newcastle, most belonging to fire insurance companies or to local big businesses who feared fires in their glass works and foundries. The Corporation itself had no precautionary measures or facilities to help with fire except for a contract with the company that supplied water to the fire plugs in the streets. Even though the number of fire insurance companies and private fire brigades had multiplied in the early years of the nineteenth century, it was only when Charles Goad began to produce his detailed fire insurance maps in the 1880s that accurate assessments of fire risk could be made.

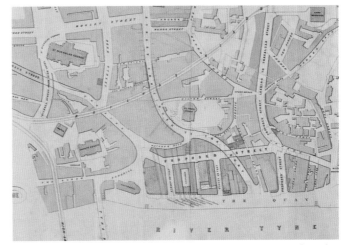

John Dobson, *Plan shewing the Intended Streets at the Quayside and approaches thereto* (1854) [SANT] Perhaps driven by grief over the loss of his son, Dobson also produced, within a remarkably short time, a series of architectural drawings of new buildings on Sandhill.

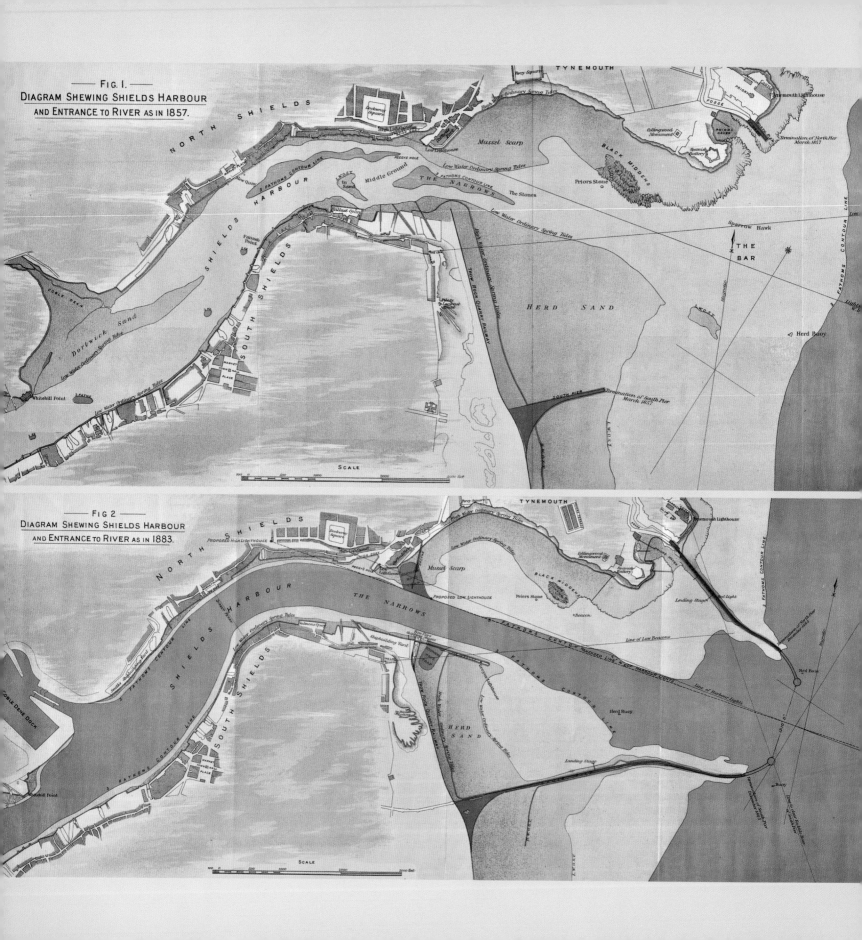

FIG. I.

DIAGRAM SHEWING SHIELDS HARBOUR AND ENTRANCE TO RIVER AS IN 1857.

FIG 2

DIAGRAM SHEWING SHIELDS HARBOUR AND ENTRANCE TO RIVER AS IN 1883.

1857

Improving the River Tyne

The condition of the Tyne was certainly not better in .
. . 1850 than it had been in 1813, at the time of Rennie's
survey – the balance of testimony . . . being that it was
actually worse.

J. Guthrie, *The River Tyne*, 1880

The contrast between the two maps of the mouth of the Tyne
provide a measure of the degree of improvement that was
necessary and then at least partially achieved in the later part
of the nineteenth century. The 1857 map clearly shows the
tortuous entrance to the river, completely open to the full force
of easterly gales. But even in fair conditions the shallow water
at the Tynemouth bar, the vast expanse of Herd Sands and the
perilous rocks of the Black Middens, as well as the Mussel
Scarp Sands revealed at half-tide, constituted significant hazards
to navigation. But this was only the entrance of the river.
Having reached the deeper but tightly constricted water at the

aptly named Narrows, it was necessary to stay to the north
but then rapidly move south away from the significant hazard
of Dortwick Sand. By contrast, the chart of 1883 shows a much
improved and straightened channel, still not without its hazards
but now with protection afforded by substantial piers, supplemented by wave-traps – groups of huge boulders angled to
break the force of heavy seas – and the removal of at least some
of the main obstructions, especially Dortwick Sand. Also
important but not visible from the map was the substantial
deepening of the river. This chapter is concerned with the story
of this immensely important transition.

Rather like the early navigation of the river itself, the
journey to achieve a safer marine channel was tortuous and
complicated. Although most aspects of Newcastle's monopoly
of the Tyne had gradually been eroded by the early nineteenth
century, the Corporation remained entirely in control of the
conservancy of the river. Only Newcastle had the power to

OPPOSITE: *Diagram shewing Shields Harbour and Entrance to the river mouth 1857 and 1883 (1857 and 1883)* [LIT & PHIL]

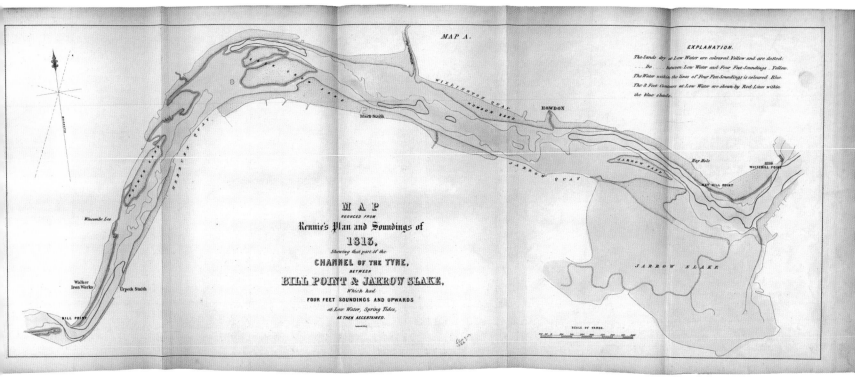

John Rennie, reduced map from *Plan and Soundings of 1813* (1813) [SANT]
The constricted and tortuous nature of the channel at low water is clear.

levy port dues with the ostensible purpose of improving the river. In reality, this money frequently found its way into financing general municipal activities. In 1774, the parliamentary candidate for Newcastle, Captain Phipps (later Lord Mulgrave) described the river as a 'cursed horsepond' and argued that the dumping of ballast was mainly due to the excessive rates charged by Newcastle, which led to ships' captains dumping ballast off the river mouth or in the river itself. The Shields' shipowners argued for the creation of a Conservancy Commission to control the river.

Goaded by the criticism, the Corporation commissioned a survey by the distinguished engineer John Rennie in 1813. His detailed survey shown here for the stretch between Bill Point and Jarrow Slake vividly shows the state of the river further upstream from its mouth. Rennie suggested that the river should be narrowed at some points to increase the scouring effect of the tide and widened at others, and that a pier should be built on the south side of the river mouth. However, virtually nothing was done, and in 1841 a further campaign was launched, accusing the Corporation of 'Misappropriation of the River Funds' and arguing that it continued to spend port dues on other matters. Support came in 1846 in the form of a Report by the Commissioners for Tidal Harbours which stated that the river was 'abandoned to itself. The improvements proposed by Mr Rennie, 30 years since, have, with the exception of a quay at Newcastle, been left unexecuted.' Representatives of North and South Shields submitted their own Bill for the creation of a Tyne Conservancy in 1848, noting that while the Corporation had received no less than £958,000 in shipping dues between 1808 and 1848, only £40,000 had been spent on direct river improvements. Eventually, Newcastle Corporation gave way and accepted the establishment of an independent Improvement Commission, initially consisting of representatives from Newcastle, Gateshead, Tynemouth, South Shields and the Admiralty.

The Tyne Improvement Act was passed in 1850, thus enabling the creation of the Tyne Improvement Commission (TIC). There was then a spurt of activity. Separate schemes to improve the river were commissioned from J.M. Rendel and W.A. Brooks in 1851, and their proposals for the mouth of the Tyne are shown on Bell's map. Rendel proposed two sets of piers, the outer ones being on the Herd Sand on the south and a smaller pier on the north, extending over the dangerous Black Middens group of rocks. The second pair would be training piers at the Narrows. He also suggested a half-tide embankment be constructed across Dortwick Sand to the west. Brooks proposed larger outer piers and more modest interventions upriver. In the light of subsequent developments, it is significant that neither report placed much emphasis on dredging the river. Nevertheless, the construction of a southern pier began in 1855 under the supervision of James Walker. The Herd Groyne, in effect the southern representative of Rendel's pair of training piers, was constructed between 1861 and 1867 and equipped with a lighthouse.

Probably the most significant achievement thus far, the Northumberland Dock, was opened in 1857 to serve the Northumberland coalfield, and the North Eastern Railway built Tyne Dock at Jarrow Slake in 1859. In the same year, John Ure, who had earlier carried out significant work on the Clyde, was appointed engineer to the Tyne Improvement Commission. He was centrally involved in changing the character of the Tyne as a commercial waterway. Fundamental to his strategy was the use of powerful bucket dredgers capable of lifting 8,000 tons per day from a depth of 34 feet. By 1863

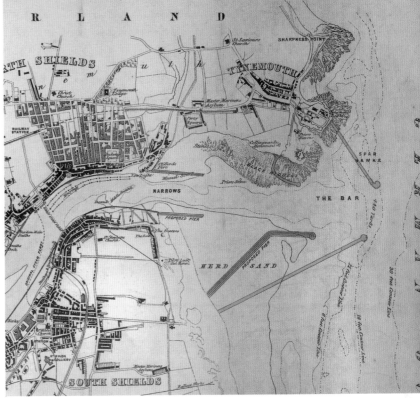

J.T.W. Bell, detail of *Plan of the River Tyne* (1852–3) [N.EST] The map shows the contrasting proposals for piers at the mouth of the river by W.A. Brooks (in red) and 'C.E.' Rendel (in purple). The latter is actually J.M. Rendel but Bell mistakenly transposed C.E. (for Civil Engineer) as his initials.

he had six dredgers working on the river below Newcastle with the aim of creating a channel 20 feet in depth. An Admiralty surveyor confirmed that Ure had deepened the channel up to Newcastle Bridge from three feet to 22 feet at low water, as shown in the long-profile of the river, adding that the Tyne

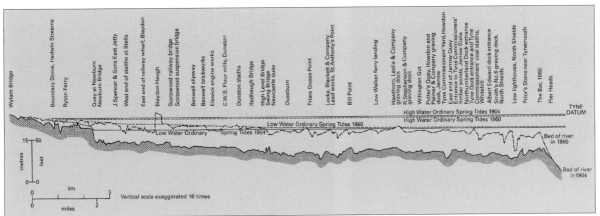

S.J. Kirkby, *Long Profile of the River Tyne* (1980) [AUTH] In 1860 the depth of water at the Bar was 6 feet 6 inches at low water. In 1879 the depth was 22 feet at low water and 37 feet at high water, spring tides.

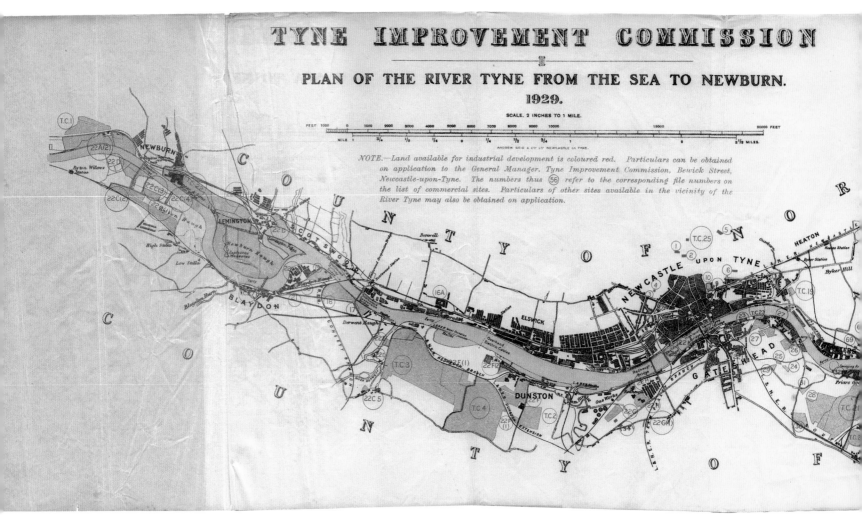

TYNE IMPROVEMENT COMMISSION

PLAN OF THE RIVER TYNE FROM THE SEA TO NEWBURN.
1929.

SCALE. 2 INCHES TO 1 MILE.

NOTE.—Land available for industrial development is coloured red. Particulars can be obtained on application to the General Manager, Tyne Improvement Commission, Bewick Street, Newcastle-upon-Tyne. The numbers thus (56) refer to the corresponding file numbers on the list of commercial sites. Particulars of other sites available in the vicinity of the River Tyne may also be obtained on application.

Tyne Improvement Commission plan of the River Tyne, showing *Land available for industrial development* (1929) [AUTH] By 1929 the TIC was responsible for a huge amount of industrial land but economic change was making much of it redundant.

represented 'the most noteworthy example of river improvements within the bounds of the United Kingdom'. Ure was also credited with initiating the Swing Bridge, engineered by Armstrong's Elswick Works, which opened the river to sea-going vessels as far west as Blaydon.

As well as deepening the river through dredging, narrowing the channel took place through the construction of stone embankments and quays. Additionally, a meander was cut off at Lemington, and between 1885 and 1887 whole islands were removed at King's Meadow and Blaydon. By 1914, 133 million tons of material had been dredged and dumped in the sea. The Commission was also active in developing infrastructure. The Albert Edward Dock, capable of taking larger vessels than the

adjacent Northumberland Dock, were added in the 1880s, along with the North Shields Fish Quay in 1886–87.

The piers at Tynemouth and South Shields were an integral part of the protection of the entrance to the river. Work on these had begun in the mid-1850s, but it was decided that longer piers were needed, leading to a change in the designs in the early 1860s. Storm damage caused further delay in 1867 and, subsequently, it was not until 1895 that the piers and accompanying lighthouses were completed. But then in 1896 a further storm breached the north pier, necessitating the complete reconstruction of the outer face which was only completed in 1915.

The relationship between the TIC and the North Eastern

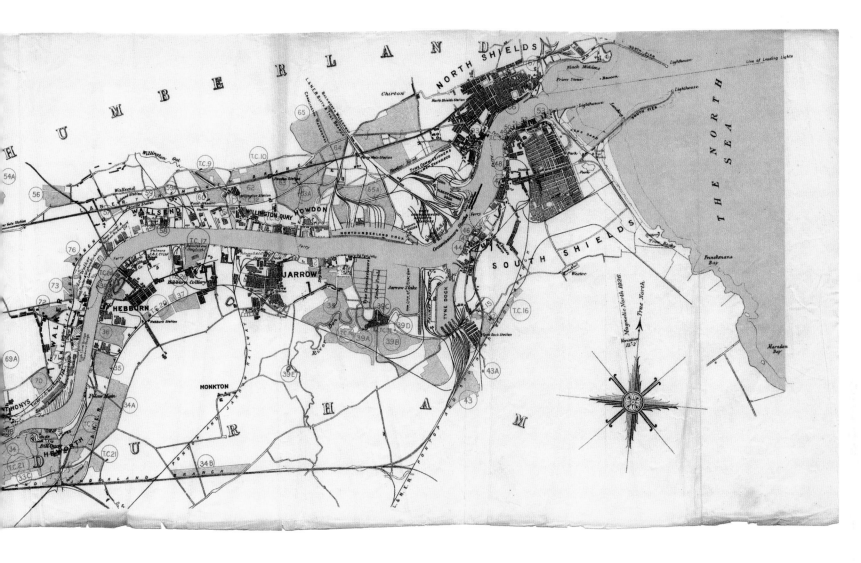

Railway (NER) Company was often fraught, although both later worked together to improve the river and the port's infrastructure. With the growth of steam-powered shipping, the NER was particularly conscious of the international demand for steam coal, and to facilitate the export trade they built the huge coal staithes at Dunston. In contrast to the previous concentration on domestic markets, by 1913 70 per cent of coal from the North East was exported for foreign bunkering and delivered to the ports by rail. Against the background of the economic difficulties of the 1920s and 1930s, the TIC began to take on a more promotional role, advertising the river and its facilities more widely and emphasising the large number of industrial sites available for occupation, as shown on the map.

In the years after the Second World War, trade on the Tyne began to decline. Some attempts were made to arrest this trend. For example, the Tyne Commission Quay, situated in front of Albert Edward Dock, was opened in 1928 and is still in use for large vessels. New coal staithes were built in 1936 at Jarrow, hoping to bring new economic activity to the town, and in 1954 at Whitehill Point, North Shields, to serve the still-operating deep mines of south-east Northumberland. Infrastructure to deal with the import of iron ore to serve the Consett Ironworks was built at Tyne Dock in 1953, and Jarrow Slake was further developed to create a terminal for the export of Nissan cars in the mid-1970s. By this time, however, the TIC had been dissolved and replaced by the Port of Tyne Authority.

Plan of
PUBLIC PARK
ON THE TOWN MOOR
and
PROMENADES
ON THE LEAZES
BY
John Hancock.
1871.

1871

Parks for the people

It was out of a varied background – public walks, pleasure gardens, residential parks, botanical or zoological gardens and arboretums – that the fully fledged public park emerged.

Mark Girouard, *The English Town*, 1990

John Hancock's beautiful map was but one of several designs produced in the 1860s and early 1870s to develop the southern area of the Town Moor and, especially, the Castle Leazes area as a public park. Other proposals and modifications came from John Fulton, the Borough Engineer, and John Laing, former steward to Sir W.G. Armstrong. Hancock was one of the founders of the Natural History Society, whose plan was influenced by eighteenth-century parkland landscape design and had no provision for games or sport. As such, it was somewhat out of step with the mood of the times. Reservoirs would be turned into landscaped water features, fringed with substantial ornamental tree planting and woodland walkways. More open promenades would be provided on the Leazes. But this was really a return to space set aside for 'promenading' by the middle classes in the eighteenth century. Few formal walkways were provided in the region, but Newcastle's Pandon Dene developed as a popular recreational venue, celebrated in the doggerel: 'When sore depressed with grief and woe / Then from a busy world I go / My mind is calm, my soul serene / Beneath the bank in Pandon Dene'.

A crucial stimulus to the provision of public parks was the passing of the 1859 Recreation Grounds Act. This empowered and encouraged local authorities to levy a rate for the laying out of public parks. As a result, public parks appeared in all

OPPOSITE: John Hancock, *Public Park on Town Moor and Promenades on the Leazes* (1871) [TWA] Hancock was invited to design Saltwell Park in Gateshead in 1875 but turned this down due to other commitments. Remarkably, the design was undertaken by Edward Kemp who had worked with Joseph Paxton to create the first purposely designed public park in the world, Birkenhead Park.

the Tyneside towns – 16 major developments in total between 1861 and 1901. The first in the region was at Windmill Hills, Gateshead, originally part of the town fields but left for common use by borough-holders after enclosure and, like Newcastle's Town Moor, a celebrated location for informal leisure and public celebrations. Several attempts to sell the land for housing development were successfully resisted, and the land was eventually conveyed to Gateshead Corporation for use as a public park in 1861.

Newcastle's first public park, Leazes Park, had similar origins, being part of the Town Moor, albeit a distinct and physically separate component. But this had now become the focus of interest in creating a public park. As well as not really meeting the requirements of a genuinely 'public' park, Hancock's scheme was considered to be too costly, partly because it included a substantial part of the Town Moor proper. A key political protagonist was the local Conservative politician Charles Frederick Hamond, a member of the Newcastle Council Parks Committee established in 1870, whose main purpose was the moral and physical improvement of the working class but under strict controls and conditions. The design of the 35-acre park was laid out by John Laing with a lake as the focal point, a skating pond and facilities for quoits, croquet and bowls. Typical of the period, however, children's play facilities were not yet provided. The park was opened in December 1873 and a little later two lodges for the head gardener and park-keeper, perimeter walls with railings, and formal entrance gates with boards displaying regulations were added. A nod towards the idea of the park as an entertainment venue took place in 1875, when a bandstand was built and became extremely popular, drawing large crowds on Sundays to enjoy free musical entertainment. Between 1878 and 1880 two smaller parks were also created out of the Town Moor area, Bull Park (later known as Exhibition Park) and Brandling Park to its east.

An earlier proposal (1863) to create a residential park at Castle Leazes would, if successful, have completely changed the subsequent trajectory of development in Newcastle. By the mid nineteenth century, suburban villa estates were increasingly fashionable, often built around or as a part of, landscaped

parks. The proposal shown here, mainly for the west side of Castle Leazes, was to be a mixed development with huge villas on the west, looking out onto the park and neighbouring lengthy rows of substantial middle-class terraces with gardens front and back.

The majority of Tyneside's major public parks, however, were created on land bequeathed or sold to councils by estate owners whose property was gradually being surrounded by urban development. Although philanthropic motives are usually mentioned in the context of the creation of Jesmond Dene, Armstrong Park and Heaton Park, constituting in R.J. Charleton's view 'the most beautiful of Newcastle's parks', the development of this linear grouping along the valley of the Ouseburn was brought about by a wider range of factors. Jesmond Dene and the area now occupied by Armstrong Park to its south had been acquired by Sir William Armstrong by the mid nineteenth century. Under the close supervision of Armstrong and his wife Margaret, Jesmond Dene was landscaped between 1850 and 1860 in romantic vein, with the retention of a ruined water mill and the importation of exotic species, as well as including waterfalls and grotto features. Armstrong allowed access to the Dene twice each week, charging a small entrance fee which was gifted to the Royal Victoria Infirmary. Immediately to its south, Armstrong Park on the wooded east bank of the Ouseburn, which had been laid out by Armstrong somewhat less adventurously than Jesmond Dene, was given to Newcastle Corporation in 1883, followed by Jesmond Dene in 1884.

Further south of Armstrong's land, Heaton Park – formerly part of the Heaton Hall estate of the White Ridley family – was acquired by the Corporation in 1879, purchased from Addison Potter mainly at the insistence of Alderman William Haswell Stephenson who, like Hamond, concerned himself with the moral and physical welfare of the working class. The Ridleys had laid out the area in the later eighteenth century,

OPPOSITE: M. and M.W. Lambert, detail of *Plan of Town Moor and Park* (1863) [TWA] The size of both villas and terraces proposed may be gauged by comparison with Leazes Terrace.

CASTLE LEAZES

Spital Tongues

Spital Tongues Ropery

Moor Cottage

Ropery Terrace

Spital Tongues Colliery

Crawney Mills

Cross Houses

North Terrace

Victoria Villas

Wardle Terrace

High Claremont Place

Claremont House

Claremont Place

Millfield House

Bull Park

Rosemount Cottage

Sharth Cottage

Brandling Vi

Warwick Place

Corporation Manure Depot

Nursery Cottage

Claremont High Terrace

S.t James Terrace

Cavalry and Infantry Barracks

Belle Grove House

APPROACH ROAD

Old Quarry

BRICK FIELD

NEW MILLS

BARRACK ROAD

Gardens

LEAZES

APPROACH ROAD

St Thomas Chapels

St Mary's Place

LEAZES PARK BRIDGE

APPROACH ROAD

ST THOMAS CRESCENT

St Thomas Crescent

RIDLEY PLACE

LOVAINE PLA

BATH ROAD

STANHOPE STREET

DIANA ST.

SWAN ST.

VICTORIA TERRACE

PITT STREET

BUCKINGHAM STREET

ELSWICK LANE

SCHOOL

PLAY GROUND

GARDENS

HALF LANE

BATH LANE

PRUDHOE STREET

PERCY STREET

ST ANDREWS STREET

GALLOWGATE

CLAYTON STREET

NORTHUMBERLAND STREET

HAYMARKET

ELDON SQUARE

BLACKETT STREET

NELSON ST.

NEW MARKETS

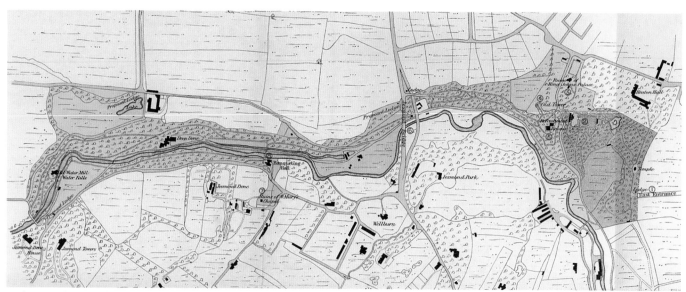

Andrew Reid, *Plan of the Armstrong Park* (1884) [ROB] A plan produced for the official opening of Armstrong Park by the Prince of Wales (later Edward VII) in August 1884.

and there had been only limited subsequent change. Conversion to a public park led to the construction of a pavilion with a balustraded terrace overlooking bowling greens, built on an area shown as a kitchen garden on an earlier map. To the south, the Ridleys had built a mock temple that was removed in the early twentieth century. A picturesque ruin is located at the north-east corner of the park, the remains of a twelfth century tower house called King John's Palace. The purchase of what became Heaton Park was thus just one component of a process that, while coming to fruition in the later 1870s, had started much earlier.

A recurrent feature of Council debates concerned the equity of provision of green open space in different parts of the city. Some argued that the Town Moor was sufficient, but others campaigned for a city-wide spread of recreational space. This particularly impacted on the western side of Newcastle where, ironically, the Natural History Society had contemplated opening a huge botanical gardens in the 1840s (*see* 1904). In 1858, over 3,000 'working men' from the west end of Newcastle had petitioned the Corporation, asking that open ground on the Elswick estate be purchased for use as a public

park, but this was met by inaction. Twenty years later, the situation was different. Labour was now organised and more effective in winning concessions and had more 'friends' in significant positions. Alderman Haswell Stephenson was active in the 1870s in agitating for the provision of public facilities for the working population such as libraries and recreational space. Along with a few like-minded councillors, Stephenson was instrumental in facilitating the Corporation's purchase of Elswick Hall and grounds in 1878. The Hall was used as a museum, and an ornamental lake, croquet lawn and promenades were developed. By the late 1870s, therefore, the east and west ends of the city had, at last, significant park provision. Middlebrook, in 1950, felt able to claim that late-nineteenth-century Newcastle 'had more open spaces reserved for public recreation than any other large town in the kingdom', but this was substantially due to the existence of the Town Moor, as Reid's 1894 map confirms.

OPPOSITE: Andrew Reid, *Green Space in Newcastle on Tyne and Gateshead* (1894) [AUTH] Although there appears to be plenty of green space on this map, its distribution raises questions about accessibility.

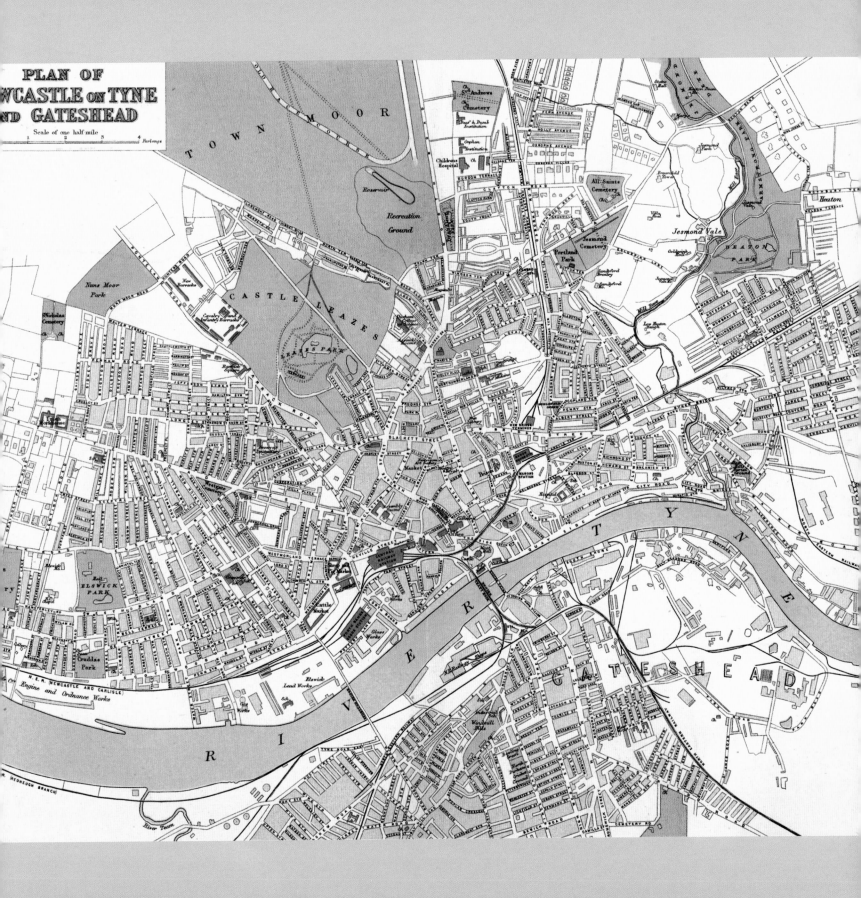

PLAN OF
WCASTLE ON TYNE
ND GATESHEAD

Scale of one half mile

1   2   3   4   Furlongs

1883.

NEWCASTLE UPON TYNE
PREVALENT ZYMOTIC DISEASES

NITIFIED BY MEDICAL PRACTITIONERS UNDER THE PROVISIONS OF
THE NEWCASTLE-UPON-TYNE IMPROVEMENT ACT. 1882.

SCARLET FEVER ⎰ × DEATHS
⎱ • Cases not fatal.

1883

The impact of a Medical Officer of Health

[A Medical Officer] 'was no more use to the town than an umbrella to a duck'.

Alderman Plummer,
Newcastle Council Reports, 1873

Despite the appalling toll conveyed by the statistics on disease, death rates and infant mortality, most officials in early- and mid-nineteenth-century Newcastle denied that the town had a serious problem. But when a Medical Officer of Health (MOH), Henry E. Armstrong, was eventually appointed in 1872, his thorough reporting – and especially his use of cartographic techniques of analysis – laid bare the facts. Armstrong saw the map as a vital tool and source of evidence, and included them in his annual reports. Unlike earlier periods, scientific evidence became a significant influence on policy development.

Newcastle had suffered more than most urban areas in the 1853 cholera epidemic, and there was much debate within the Council on the causes and possible remedies. Taking water from the Tyne, while a major contributing factor, became a convenient excuse for limited or even lack of action over other factors, especially relating to housing conditions. In 1861, the influential periodical *The Builder* wrote in excoriating terms: '. . . with their mercantile success and gain, and eagerness still further to enrich themselves, the people of Newcastle have . . . utterly neglected all provision for their health'. Contemporary statistics bear out this assertion. The average annual death rate was 25.7 per 1,000 in 1841–48, but had climbed to 32.1 in 1866, then the third highest rate in the country.

Yet inertia prevailed for years. In the 1853 cholera outbreak, the Council waited for two weeks before meeting to discuss it, leading the Chief Inspector from the Central Board of Health to be extremely critical of their failure to enforce sanitary regulations. Newcastle, along with the other Tyneside

OPPOSITE: Henry E. Armstrong, *Scarlet Fever, 1883 epidemic* (1883) [NCL] Typical example of Armstrong's disease mapping.

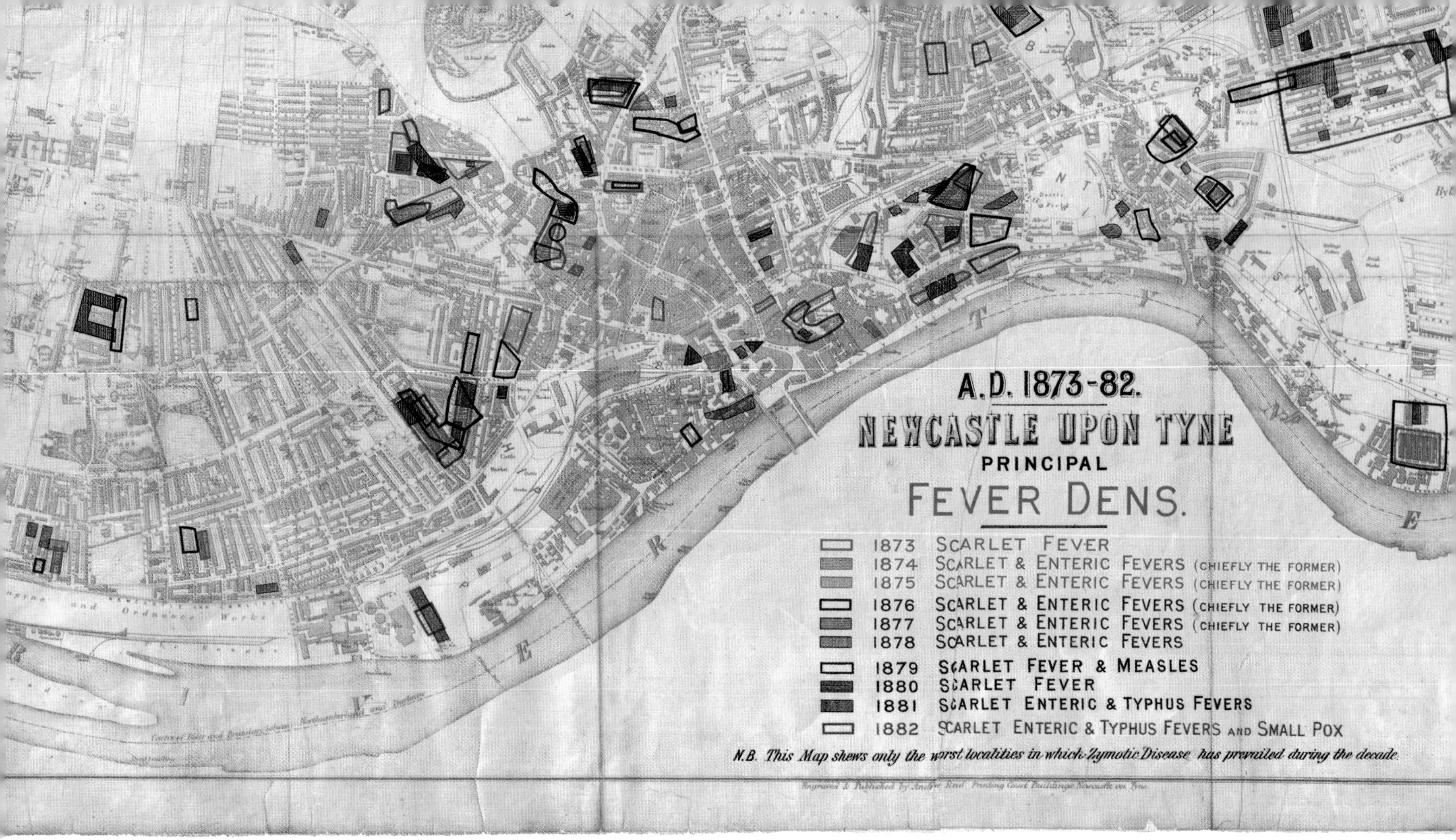
A.D. 1873-82.
NEWCASTLE UPON TYNE
PRINCIPAL
FEVER DENS.
1873 SCARLET FEVER
1874 SCARLET & ENTERIC FEVERS (CHIEFLY THE FORMER)
1875 SCARLET & ENTERIC FEVERS (CHIEFLY THE FORMER)
1876 SCARLET & ENTERIC FEVERS (CHIEFLY THE FORMER)
1877 SCARLET & ENTERIC FEVERS (CHIEFLY THE FORMER)
1878 SCARLET & ENTERIC FEVERS
1879 SCARLET FEVER & MEASLES
1880 SCARLET FEVER
1881 SCARLET ENTERIC & TYPHUS FEVERS
1882 SCARLET ENTERIC & TYPHUS FEVERS AND SMALL POX
N.B. This Map shews only the worst localities in which Zymotic Disease has prevailed during the decade.

Henry E. Armstrong, *Principal Fever Dens* (1873–82) [TWA] Only the 'worst localities' are mapped. There were clearly plenty of them.

boroughs, appointed an MOH only when forced to do so. The prevalent view within the Council was that, unlike commercial and industrial growth, sanitary action was not its concern. Remarkably, this view was shared by such leading medical practitioners on the Council as Sir John Fife and T.E. Headlam, to both of whom the concept of *public* health was a foreign one. However, electoral reform led to significant change in the composition of the Council in the 1850s, including the election as councillor of a very different type of medical practitioner, epitomised by Dr William Newton, who was a union medical officer in a very poor district and dedicated to the idea of sanitary reform being a municipal responsibility and an imperative. Although never a councillor, another profoundly influential doctor was J.H. Rutherford, preacher, educationalist and author of an important sanitary report in 1866. Probably even more significant was the increased pressure for sanitary reform exerted by intellectual liberal industrialists such as Isaac Lowthian Bell and Joseph Cowen Jr.

However, an influential group within the Council argued that Newcastle did not need an MOH. Alderman Gregson simply did not believe the official statistics on death rates: the medical men he had spoken to said 'they had almost no sickness' – this despite a death rate of 30.1 per 1,000 people in 1873 when the average death rate for the 21 largest towns in England was 24.3, and the death rate for children under one year of age being 220 per 1,000. But national legislation gave the Council no choice but to make such an appointment. Armstrong retained the post for almost 40 years until his retirement in 1912. Aware of the tendency towards denial of the facts within certain sections of the Council (notably Alderman Plummer, as quoted above, who was presumably unconcerned about prenatal mortality, suggesting that the figures were a scam: 'To get the fees from the burial clubs, the children were returned as having died from some plausible disease, whereas they were never born alive at all.') Armstrong provided a large volume of cartographic and statistical evidence in his annual reports. His map of the scarlet fever outbreak in 1883 is typical of his work. Depending on their severity, from 1873 Armstrong mapped the annual incidence of smallpox, typhoid, enteric fever, measles, whooping cough and diarrhoea, in addition to scarlet fever.

A revelation for the Council was that, in addition to 'expected' high incidence of disease in places like Ouseburn with its proximity to noxious industry such as lead works, Armstrong showed that more recently built areas such as 'Gibson Town' and east Shieldfield and, even more shockingly, the new peripheral districts in Byker and Elswick, also had high concentrations of zymotic diseases. Armstrong was eager to demonstrate the links between various diseases and living conditions and their concentration in particular areas. This was the basis of a campaign to identify and isolate 'fever dens' where, as shown on the map, there was a year-on-year overlap of the incidence of disease. But he was also aware of the role of such areas in disseminating disease: 'From these dens come charwomen, errand boys, newspaper vendors, seekers after Dispensary letters and mendicants.'

Henry Armstrong became an extremely important figure in the improvement of public health in Newcastle in the latter part of the nineteenth century and early part of the twentieth. By 1901, Newcastle's death rate had reduced to 20.1 per 1,000 population. Although much of the improvement was a product of the enforcement of national legislation, Armstrong was proactive in dealing with issues such as types of human waste disposal, public nuisances such as noxious trades, urban dairies and the quality of food and drink retailed. Typical of his period and class, he also pointed out the role of social characteristics, especially the negative impact of unskilled manual labourers and, even more, Irish immigrants 'whose habits and love for over-crowding render them more usually liable to sickness'. But, with the exception of cellar dwellings, he was extremely reluctant to act in the field of the housing market. Sidney and

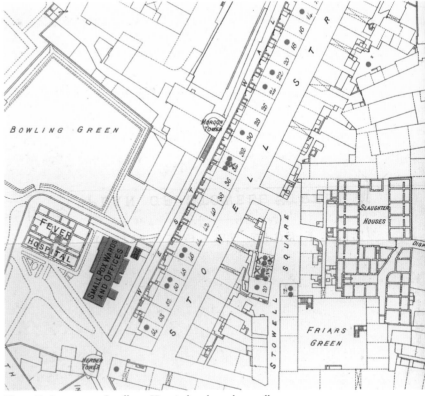

Henry E. Armstrong, *Smallpox Hospital and nearby smallpox cases, Stowell Street* (1881) [NCL] The smallpox 'hospital' was a 56-bed wooden addition to the Fever Hospital.

Beatrice Webb, who interviewed him as part of their research into the delivery of public administration in English towns in the early twentieth century, noted his unwillingness to become involved in housing districts where leading councillors were significant property owners.

Nevertheless, Armstrong was a most capable and innovative MOH, actively involved in researching the sources of particular health problems and in providing a detailed annual description and explanation of the incidence of disease, down to the level of individual streets. Although careful to cite other factors, his cartographic demonstration of the possible role of the specialist smallpox ward adjacent to Stowell Street in exacerbating smallpox within that district was instrumental in its removal to the less inhabited area of the Town Moor.

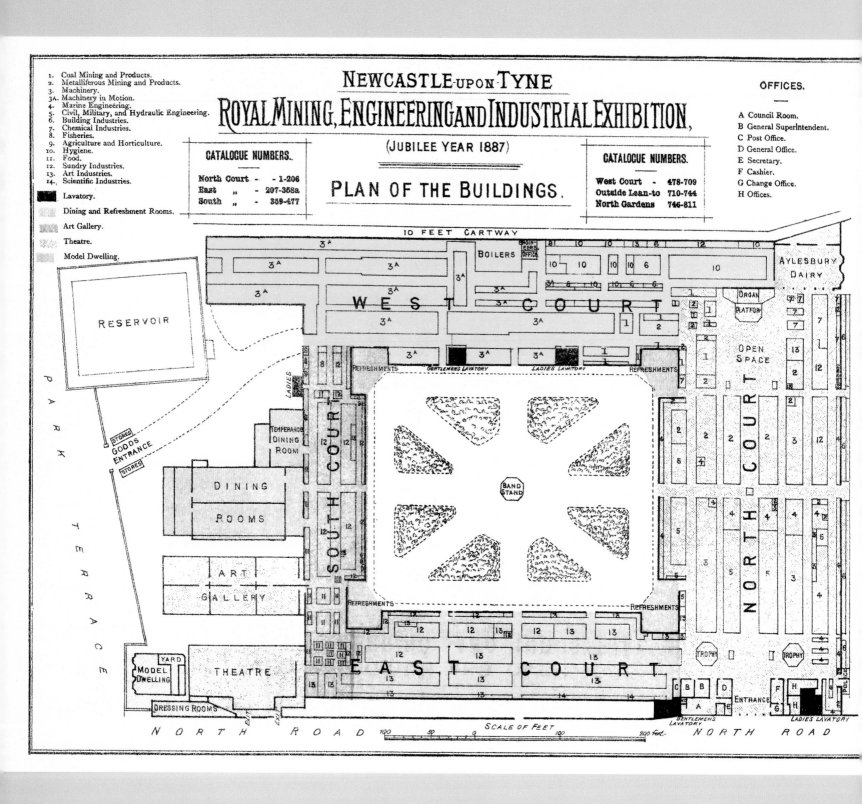

NEWCASTLE-upon-TYNE
ROYAL MINING, ENGINEERING and INDUSTRIAL EXHIBITION,
(JUBILEE YEAR 1887)
PLAN OF THE BUILDINGS.

1. Coal Mining and Products.
2. Metalliferous Mining and Products.
3. Machinery.
3A. Machinery in Motion.
4. Marine Engineering.
5. Civil, Military, and Hydraulic Engineering.
6. Building Industries.
7. Chemical Industries.
8. Fisheries.
9. Agriculture and Horticulture.
10. Hygiene.
11. Food.
12. Sundry Industries.
13. Art Industries.
14. Scientific Industries.

Lavatory.
Dining and Refreshment Rooms.
Art Gallery.
Theatre.
Model Dwelling.

CATALOGUE NUMBERS.

North Court - 1-206
East „ - 207-358a
South „ - 359-477

CATALOGUE NUMBERS.

West Court - 478-709
Outside Lean-to 710-744
North Gardens 746-811

OFFICES.

A Council Room.
B General Superintendent.
C Post Office.
D General Office.
E Secretary.
F Cashier.
G Change Office.
H Offices.

10 FEET CARTWAY

WEST COURT

RESERVOIR

BOILERS

ENGINEERS OFFICE

AYLESBURY DAIRY

ORGAN PLATFORM

OPEN SPACE

NORTH COURT

PARK TERRACE

GOODS ENTRANCE

STORES

LADIES

SOUTH COURT

TEMPERANCE DINING ROOM

DINING ROOMS

ART GALLERY

REFRESHMENTS

GENTLEMENS LAVATORY

LADIES LAVATORY

REFRESHMENTS

BAND STAND

REFRESHMENTS

REFRESHMENTS

TROPHY

TROPHY

MODEL DWELLING

YARD

THEATRE

EAST COURT

DRESSING ROOMS

ENTRANCE

POLICE

C B B D A F H G H

GENTLEMENS LAVATORY

LADIES LAVATORY

NORTH ROAD

SCALE OF FEET

100 50 0 100 200 feet.

NORTH ROAD

1887

Exhibitions: agriculture, industry and fun

Upon a bleak Northumbrian moor
Behold a palace raised. Behold it filled
With all that fingers fashion, deftly skilled,
With all that strongest fibred brains have willed . . .

Thomas Hodgkin, 1887

Hodgkin's 'bleak Northumbrian moor' was actually the Town Moor. His typically Victorian ode was specially written to celebrate the opening of the 1887 Golden Jubilee Exhibition there. Despite the semi-rural environment of the Town Moor, its south-east corner was about to celebrate a history that recognised a very different image of Newcastle. The area became known as Exhibition Park, famous for an essentially 'urban-industrial' representation of the city arising from a sequence of major exhibitions.

However, the template for large-scale exhibitions had been laid down 20 years earlier by the 1864 Royal Agricultural Society (RAS) Exhibition, when the region's elite had been flattered to be given the opportunity of mounting this national event. In 1887 the RAS returned to Newcastle and its Town Moor as part of the celebrations for Queen Victoria's Golden Jubilee. But that year the main event, attended by over two million visitors between May and October, was the Royal Mining, Engineering and Industrial Exhibition. Its title signalled the real purpose of the event. Although ostensibly celebrating the Queen's 50 years on the throne, the main message concerned the promotion of the industrial economy of the city and its region. Even by Victorian standards, the size of the Exhibition's organising committee was remarkable, with no fewer than 12 subcommittees, possibly designed to ensure that none of the Tyneside elite felt excluded but also reflecting the strength of enthusiasm for the event, which was itself a product of a

OPPOSITE: *Plan of Royal Mining, Engineering and Industrial Exhibition, 1887* (1887) [AUTH]
The bandstand in the centre is the only physical remnant of the 1887 Exhibition.

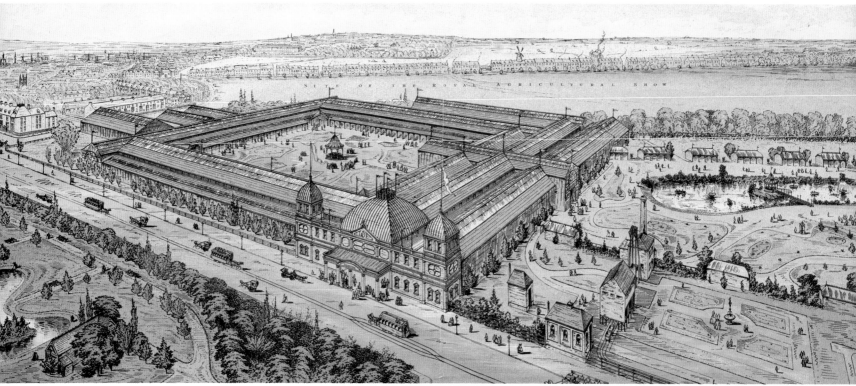

Bird's-Eye View: Royal Mining, Engineering and Industrial Exhibition (International and Colonial) (1887) [TWM] The 'working coal mine' is clearly visible to the bottom right of the image but the replica of the old Tyne Bridge (destroyed 1771) had not been constructed at the time of this drawing.

combination of factors including a desire to demonstrate to the world the marvels of modern industrial Tyneside and the genuine surge of patriotism evident in Thomas Hodgkin's ode.

As shown on the 1887 Exhibition plan, the layout was broadly rectangular. Altogether, there were 811 catalogued displays plus a theatre, art gallery and model dwelling exhibition, all surrounded by gardens. The north court of the exhibition was significantly larger than the others and featured manufacturing industry, particularly metal industries, marine engineering, hydraulics and mining, although the space devoted to the latter was surprisingly small. On the west, 'machinery in motion' was featured, rather oddly juxtaposed with 'hygiene' exhibits. Food industries were located on the south, with the east mainly accommodating 'arts-related' industries. The central area was dominated by gardens and a bandstand which

still exists. The most remarkable features of the Exhibition were the historical reconstruction of the old Tyne Bridge, complete with buildings, and the reconstruction of a working coal mine, to be seen on the near right of the bird's-eye view (the 'model' of the old Tyne Bridge had not been completed at the time of this perspective view).

Contrary to the region's effusive and rather self-congratulatory response, the *Builder* magazine published a detailed and somewhat condescending review, unfavourably comparing the Exhibition to that in Manchester. Commenting on the militaristic character of much of the Exhibition, the report, remarkably for a periodical of this era, also drew a regretful parallel between the symbolism represented by the intimidating impact of the great Castle Keep and Armstrong's 111-ton 'monster' gun on display at the Exhibition. Furthermore,

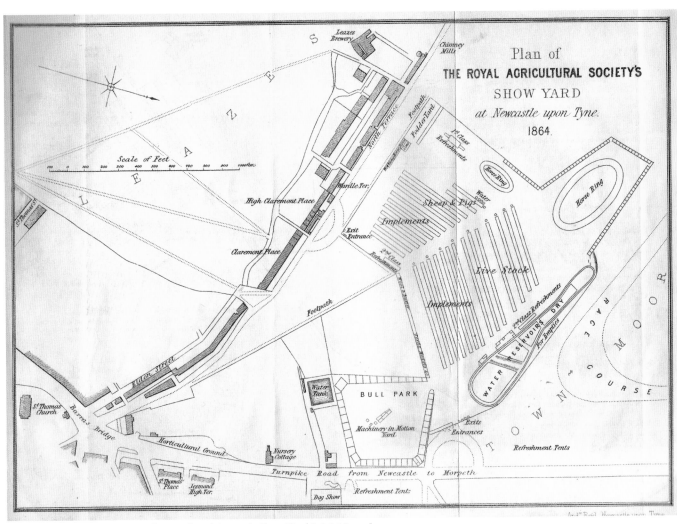

Andrew Reid, *Plan of the Royal Agricultural Society's Show Yard* (1864) [SANT]
Note the segregation of '1st Class refreshments' and '2nd Class refreshments'.

despite the 'official' interpretation of the event as a celebration of industrial technology, the final report noted that the most popular features were the 'toboggan slide' and the 'switchback railway'. As we have noted, the RAS Exhibition was held for one week during this Jubilee, with the two exhibitions cohabiting successfully and, far from detracting from each other, proving to be complementary.

Although the term 'Exhibition Park' dates from the Royal Jubilee exhibition of 1887, this area had been used for 'exhibi-

tions' or similar events for many years before this. Prominent amongst these and somewhat incongruously given its location close to the centre of Newcastle, the most significant were agricultural shows. But the display of agricultural progress fitted well with the semi-rural environment that the Freemen of the city wished to maintain. The Royal Agricultural Society (RAS) had held its annual show in Newcastle in 1846, but its 1864 show was considerably more significant.

The RAS contacted Newcastle Corporation, indicating they

Above and detail opposite. V.L. Danvers, *North East Coast Exhibition Poster* (1929) [NCL] Danvers also illustrated 'The River Tyne: a Great Industrial River', a promotional booklet issued by the Tyneside Development Conference, *c.*1930.

were considering several locations in the north to hold their 1864 Exhibition. Their 'wish list' was a long one, including a request that the Corporation guarantee a sum to cover the Society's expenses, free use of a portion of the Town Moor for a show yard, the levelling and draining of the site, the power to erect temporary buildings and free use of the Town Hall for the transaction of business. Only Alderman Blackwell dared to raise a tentative objection, asking, 'What do they give us for all this?', to which John Clayton, the all-powerful town clerk replied, 'They give you the honour of their company.' The Corporation pledged to guarantee a sum of £2,000 to be raised by subscription. In fact, such was the interest that over £5,000 was raised. This apparent enthusiasm reflected the fact that the 'landed interest' retained considerable power and influence within the city at this time. The Newcastle Farmer's Club (established in the 1840s) was a significant local institution

with no less than five MPs on its committee in 1865, including Sir Matthew White Ridley and Henry George Liddell (2nd Earl Ravensworth), both substantial landowners but also with major interests in the town.

There was a major contrast between the Jubilee Exhibition of 1887 and the city's next major exhibition, the North East Coast Exhibition of 1929, displayed on the accompanying image. While the former was a genuine celebration of the region's role and achievement, the latter was, in many ways, a rather desperate attempt to counter the economic depression of the interwar years. Despite its intention to regenerate industry within the region, the 1929 Exhibition was as much about fun and entertainment as about promoting the region. Considerably larger than the 1887 Exhibition, although similarly temporary in its infrastructure, the main buildings were the Palace of Engineering, the Palace of Industry, the Palace of Arts and the government-sponsored Empire Marketing Board Pavilion. But large areas were devoted to the sports stadium holding 20,000 people and to a funfair with modern rides, both displayed on the top right of the poster. Most of the artwork for the exhibition's advertising was executed by local commercial artist Paul J. Brown but the poster shown here was the work of Verney L. Danvers, one of the best-known modern graphic designers at this time, famous for his posters for London Transport. Modernity was also signalled by the almost unprecedented role of the BBC in broadcasting concerts from the Exhibition to the rest of the nation and the provision of substantial car-parking space. The event attracted over four million visitors, double that of 1887, and in many ways reflected the confusion of the times: was it about celebrating past economic successes or was it about the future and the showcasing of new industries and commercial activities? In retrospect, it was the latter that was the victor, and the most memorable impact in the folk memory was made by the modern buildings, the role played by electricity, the displays of new consumer items, the nightly fireworks and, quite simply, the people having a good time and being cheered up.

1890

Newcastle's Town Hall saga

I noticed the crying need for a town hall worthy of a city with such records, and with such immense latent possibilities; a town hall which would provide the facilities which at present in Newcastle are conspicuous for their absence.

J. Douglas Mitchell, *The Architect's Journal*, 1 July 1925

Nearly a century of bitter argument about a suitable location for a new Town Hall was laid to rest when, on the cold blustery day of 14 November 1968, Norway's King Olav V formally opened Newcastle's new Civic Centre at Barras Bridge. Few topics had generated as much heated debate within the Council and few had revealed as much about the vested interests, ideological positions and personal characteristics represented within the Council.

From the 1850s until the opening of the Civic Centre, the Town Hall had been located in the former medieval market area facing the Cathedral. The foundation stone for the building was laid in 1855, and the dual-purpose Town Hall and Corn Exchange building opened in 1858. The siting of this development was controversial. Richard Grainger argued that it would block the view of St Nicholas with its dramatic lantern spire and would destroy the possibility of having a spectacular street curving down the Bigg Market with views of the Black Gate and Castle Keep. He even offered his Central Exchange as a home for the Corn Exchange, but all to no avail.

Furthermore, the new Town Hall had not been open long before its inadequacies began to emerge. By the 1880s the building was being condemned as overcrowded and unhygienic, with inadequate conveniences and a proliferation of rodents (ironically, the worst affected area being the Health Department). A Special Committee was established in 1890, charged

OPPOSITE: Property Office, Town Hall, *Plan shewing sites proposed for New Town Hall* (1890) [SANT]
In the absence of clear criteria, Town Hall sites were identified mainly on the grounds of likely availability.

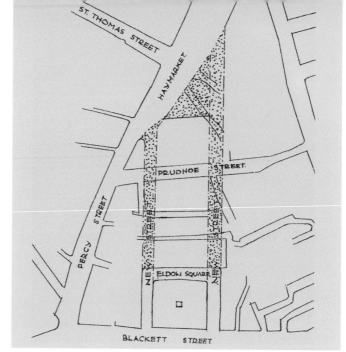

Ordnance Survey, *Detail of Town Hall and Corn Market Buildings* (1858–60) [SANT] The new Town Hall of 1855 (not completed until 1863) had a strategic but severely constricted location.

J. Douglas Mitchell, *Proposed Eldon Square site for new Town Hall* (1924) [NCL] Mitchell was a prominent member of the Newcastle upon Tyne Society which continually pressed the Council to engage more seriously with civic development.

not only with finding a suitable site but also with establishing 'the selling value of the present municipal offices'. The Committee's report included the prospect of making suitable alterations to the existing Town Hall, together with suggestions of no fewer than 14 potential new sites. However, these were rapidly narrowed down to the five locations shown on the map and, after fierce debate to just three: the north side of Eldon Square; Singleton House on the corner of Northumberland Street and Bath Road (now Northumberland Road); and the Assembly Rooms site on Westgate Road. After prolonged debate in the Council, a six-month adjournment was agreed and, on returning to the issue in October 1891, a new consensus had emerged in favour of improving the existing building!

If the Council felt they had resolved the issue, however, others did not agree. New voices began to be heard, especially from the modern profession of town planning. The architectural firm of Cackett, Burns Dick & Mackellar emerged as a powerful force. J.T. Cackett (1860–1928), president elect of the Northern Architectural Association, took advantage in his inaugural lecture (1905) on 'Newcastle Improvements' to remind his audience of the inevitability of change in the twentieth century, highlighting the advent of the motor car and

its impact on the city centre. He further noted that the purchase of property associated with the recent eastern extension of Market Street 'provides a site most suitable for a Town Hall, municipal buildings, and Police and Assize Courts'. Although the Town Hall was not built there, Cackett's plan was instrumental in the construction of the police station, magistrates' court and fire station on that site. The latter buildings were partly the work of Robert Burns Dick (1868–1954), then a junior partner of Cackett and, although nowadays barely recognised as such, probably the most significant figure in influencing Newcastle's built environment in the interwar period.

Burns Dick's plan for the city centre (*see* 1924) featured a new Civic Centre at the head of a wide boulevard leading from Barras Bridge. Although, like many others relating to a new Town Hall, this scheme was never implemented, Burns Dick succeeded in turning thoughts to the north of the city centre. Remarkably, however, the interwar period witnessed an almost exact repeat of Council debates of the 1890s. By 1927, although it was confirmed that the Burns Dick plan was perceived as too grandiose, his instinct for a meaningful site for a new Town Hall had won over many in the Council. A site fronting the North Road between Jesmond Road and

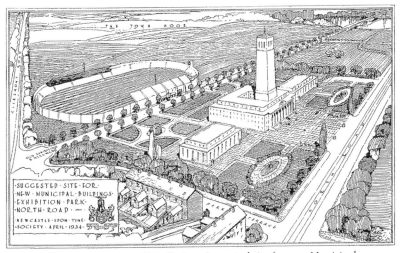
Newcastle upon Tyne Society, *Suggested site for new Municipal Buildings, Exhibition Park, North Road* (1934) [AUTH] This was the site of the 1887 Jubilee Exhibition.

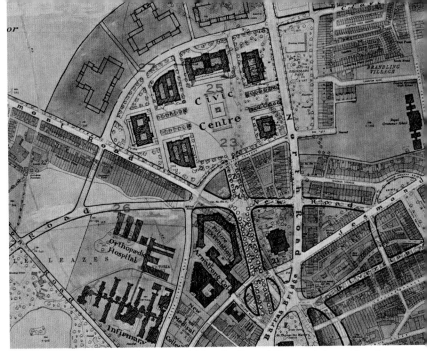
Robert Burns Dick, detail of *Burns Dick Plan for Central Newcastle* (1924) [NCL] Design influenced by the 'City Beautiful' movement, developed in North America. Monumental grandeur in civic buildings and their layout, including tree-lined avenues, were thought to enhance civic culture.

Sandyford Road was now favoured, although reactionaries rolled out their usual arguments for yet another refurbishment of the existing Town Hall. Indecision was rife and, by January 1929, the Council took the then unusual step of employing a consultant expert, H.V. Lanchester RIBA, to advise them on the Town Hall question. Lanchester's report clearly endorsed the Burns Dick plan of 1924 with his recommendation of what he called the 'Town Moor' site. Yet the committee appointed to consider the question resolved to concentrate on the Eldon Square site. In an uncanny repetition of previous debates, the ensuing furore allowed the question of modernising the existing Town Hall to be raised yet again, particularly by Alderman Easten, a local builder. The 'choice', within the Council at least, now seemed to be between the Eldon Square site or renovation of the existing site.

But by this time Burns Dick had become a leading member of the Newcastle upon Tyne Society, a pressure group that saw its role as trying to persuade the City Council to adopt modern ideas of town planning and environmental management. In 1930 the Society intervened, pointing out that the expert consultant had recommended a site on Exhibition Park and putting forward a slightly adapted version of Burns Dick's plan.

This intervention appears to have been decisive in getting the Council to agree that a site to the north of the city centre was the most sensible choice. By spring 1930 a site at Barras Bridge, known as the St Thomas' Church grounds, began to be seriously considered, with the choice of site now being between this, Exhibition Park and Eldon Square. The latter retained the support of many in the Council (mainly those with city-centre business interests), but the continued inability to reach a decision caused the issue to be postponed, a delay further exacerbated by the onset of the Second World War.

The pressure of the Newcastle upon Tyne Society had some influence on the first formal town planning document for the city, Percy Parr's 1945 *Plan for Newcastle*. In many respects, it was an even more ambitious version of the Burns Dick plan of 1924, especially in relation to new roads in and around the city centre. But, fascinatingly, a centrepiece of this 1945 plan was yet again a new Town Hall which, albeit with a completely revolutionised road pattern (actually, never constructed) at the north end of the city centre, was located just to the north of St Mary's Place and the church of St Thomas the Martyr – the actual site of the Civic Centre opened officially by King Olav V in 1968.

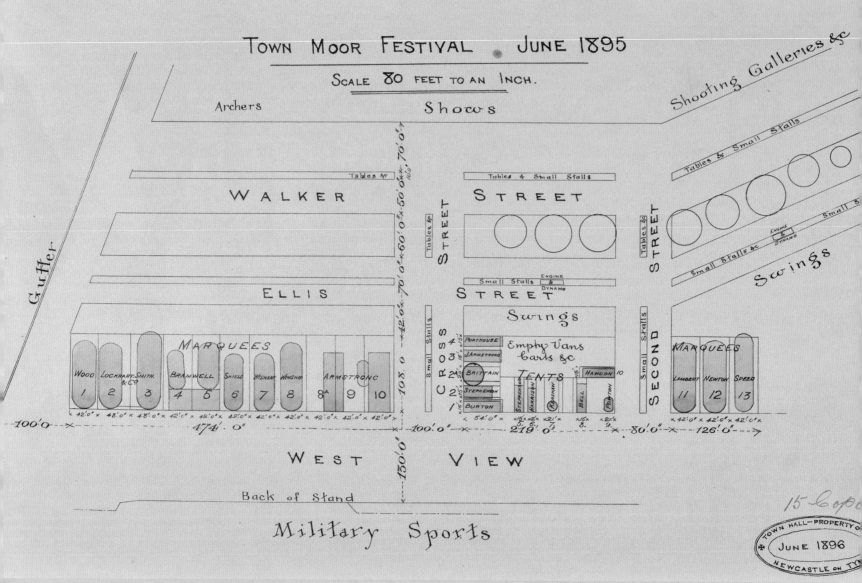

TOWN MOOR FESTIVAL · JUNE 1895

SCALE 80 FEET TO AN INCH.

Archers · Shows · Shooting Galleries &c

WALKER · STREET

Tables &c · Tables & Small Stalls · Tables & Small Stalls

ELLIS · STREET

MARQUEES

Small Stalls · Engine & Dynamo · Small Stalls &c

Swings

Small Stalls · Empty Vans Carts &c · Swings

WOOD 1 · LOCKHART SMITH & Co 2 · 3 · BRAMWELL 4 · 5 · SHIELD 6 · WISHART 7 · WINSHIP 8 · ARMSTRONG 8A · 9 · 10

CROSS · SECOND · STREET

PORTHOUSE 4 · J ARMSTRONG 3 · BRITTAIN 2 · STEPHENSON · BURTON 1

STEPHENSON 5 · HARRISON 6 · BOOZMAN 7 · BELL 8 · HAWDON 10 · PEXTON 9

TENTS

MARQUEES · LAMBERT 11 · NEWTON 12 · SPEED 13

WEST VIEW

Back of Stand

Military Sports

Gutter

x 42'0" x 48'0" x 48'0" x 42'0" x 42'0" x 42'0" x 42'0" x 42'0" x 42'0" x 42'0" x 42'0" x
471'·0"
100'0"
100'0"
150'0"
219'·0"
80'0"
126'·0"
x 54'0" x
x 42'0" x 42'0" x 42'0" x
108'·0 — 42'0" — 70'0" — 60'0" — 50'0" x 70'0"

1895

The growth of the Hoppings

It is a necessary qualification for all 'good Geordies' to have visited the event at least once in their lifetime . . . known to the majority of them as . . . 'The Town Moor Hoppings'.

F. Barron, *The Town Moor Hoppings*, 1984

J.T. Brockett, the Victorian Tyneside antiquarian, claimed that the word 'hoppin' [sic] referred to a country wake or rural fair characterised by the dancing of young people. But by the early nineteenth century the term was being used in industrial areas in relation to holiday gatherings such as at the village of Wylam in the Tyne Valley and at Windmill Hills in Gateshead. Within Newcastle in the mid nineteenth century, the 'hoppings' held on the Forth and Marlborough Crescent caused considerable nuisance and concern to the Corporation. In modern times, the term has come to refer to a very specific event, and most

Tynesiders would now associate it exclusively with the annual fair held towards the end of June on Newcastle's Town Moor. The event is often claimed to be Europe's largest travelling fair occupying 40 acres of the Town Moor, with over a million visitors (many of them repeaters) being entertained by 600 members of travelling-show families.

Although the atmosphere of the event is one of spontaneity, its origins and current management reflect the desire to exercise a degree of control over its use of space. The Hoppings owes its existence to the City Council's desire to curtail the more rowdy and alcoholic elements associated with Newcastle's annual Race Week. From 1721 to 1881 this much-celebrated occasion gradually extended in scale and range of activities beyond horse racing, attracting up to 50,000 visitors. Like many race meetings, the spectators ranged across the social spectrum, and the races became an important part of the

OPPOSITE: Property Office, Town Hall, *Town Moor Festival, June 1895* (1895) [SANT]
The 'festival' was clearly not a spontaneous event but one that was strictly controlled.

Lamberts, *Programme of the Temperance Festival, June 1882* (1882)
[SANT] Note 'The Publican's Finished Work' in the bottom right.

decided to hold a Temperance Festival on the Town Moor. This built on the fact that over the previous decade various temperance groups around Tyneside had established annual public gatherings, and momentum was growing to combine them into a much larger single event. In March 1882, a local newspaper carried a suggestion that this could be held on the Town Moor at the time that Race Week would have taken place and be combined with a range of entertainment and activities. Support was forthcoming from the Newcastle Temperance Society, the Gospel Society and the YMCA. A number of showmen attended this 'festival', which turned out to be extremely successful, with over 150,000 visitors. It was repeated the following year, was an even bigger success, and thereafter became an annual event. The plan of 1895 shows the well-ordered regularity of the layout. The first festivals lasted for two days and had a focus on competitive games. The newly popular bicycle figured prominently in race events, along with football, archery and shooting.

But this success carried with it the challenges of ensuring a safe and efficient event. The freehold of the land where the fair takes place is owned by the Council, but the Freemen of the City have a number of rights and privileges which had to be protected. The Hoppings could therefore not take place without the agreement of both parties. Over the years this has led to problems. In 1912 very heavy rainfall led to waterlogged ground and the Freemen claimed that showmen had badly damaged their pasture. They attempted to prevent the 1913 fair, although a court case found in favour of the showmen; in 1914 they were granted an injunction. No fair was then held on the Town Moor until 1924, although a smaller version of the Hoppings was held in Jesmond Vale during each year of the First World War. The organisation of the Hoppings remains complex. The Northern Showmen Syndicate retains the lease of the Town Moor, which they rent from Newcastle Council for the duration of the event, and therefore has an important voice in the organisation of the show. Early each year the showmen's trade journal, *The World's Fair*, publishes the deadline for applications for stands within the show, although the larger rides and long-established show families have

county gentry's social calendar. A two-storey grandstand was built on the northern side of the Town Moor, also functioning as a public tavern throughout the summer.

However, in 1881 the race meeting was transferred to Gosforth Park, reasserting the Town Moor as essentially the space of the Freemen of Newcastle, albeit with some concessions for general access. In the following year, however, it was customary rights, locating on the same patch each year.

For many Tynesiders from the 1880s to the 1930s, this was their annual holiday. Despite the recession of the interwar years, the size of the Hoppings site grew. The plan of 1895 shows that the fair was laid out at the north of the Moor backing onto the grandstand. Thirteen large marquees were rented out to nine different showmen or organisations, with a further 11 smaller rented shelters. These were complemented by several rows of tables and small stalls. There was an emphasis on children with the provision of swings, with many of the competitions being organised for children and with a free tea being provided for children from poor neighbourhoods on one of the days. In addition, for several decades prominent features included sporting events, military displays and musical performances. The opening ceremony was a formal event with exhortations on the evils of alcohol and the benefits of teetotalism. The printed programme of the very first festival captures the spirit and purpose of the event.

Innovations such as cinematographic booths appeared at the turn of the century and many Geordies saw their first moving pictures at the Hoppings. By 1925 four rows of main shows were laid out, still based on the north-east part of the Moor, but the nature of entertainment had changed considerably. The rather earnest early presentations had given way to both a greater fairground atmosphere, including grotesques and freak shows, and a circus-like tone with animal shows, including Professor Testo's Flea Circus. Strongman competitions, wrestling and boxing were annual regulars. By 1935 the ground taken had increased considerably, extending all the way from Grandstand Road to Exhibition Park with four long rows of main attractions. By this time the tradition of a travelling fairground was well established.

In the latter part of the twentieth century, the holiday reached new heights of popularity, being extended to nine days from Friday through to the Saturday of the following week inclusive. The record attendance of some 1,250,000 was achieved in 1979 and a total of 26,072 cars paid the parking fee. Not only does the Hoppings fair demonstrate continuity over more than a century, but the show families also reveal a

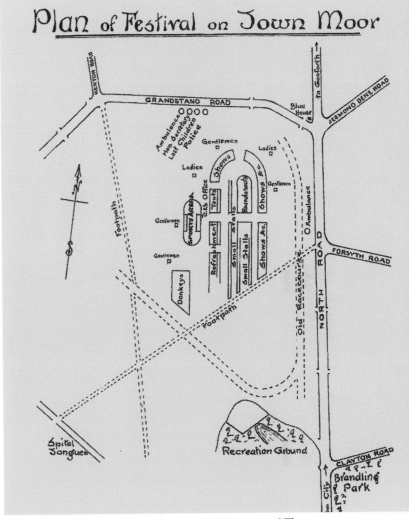

Sketch plan of Festival on Town Moor (1926) [BARON, 1984] The 'Sports Arena' and 'Donkey' area indicate a very different 'festival' than the contemporary Hoppings.

great capacity to adapt to changing times, with the older forms of amusement being replaced by bigger and more spectacular entertainments, including the 'Wall of Death', 'Big Wheel', 'Dive Bomber', dodgems, ghost trains and fun houses. There is no doubt that the event has now undergone a complete transformation in character from 'Temperance Festival' to 'Fun Fair'.

1896

Music for the masses

Newcastle could almost claim to be the home of the Music Hall. It flourished here as nowhere else.

F. Graham, *Historic Newcastle*, 1976

The Goad fire-insurance plan of the Cloth Market area of central Newcastle in 1896 shows one of the town's historically premier – and persisting – social leisure districts. Justly renowned for their detail, the main impetus for producing maps at this large scale came from concern over the risk of fire. Building materials, room arrangements, type of roof and other features were mapped, including the use of different parts of buildings. The mixture of different land uses is evident, but a prominent feature is the density of public houses and related places of entertainment such as music halls. In 1882 the area of Bigg, Groat and Cloth Markets provided sufficient thirst to warrant the location of 23 pubs and four beer houses. The boundary between public house and music hall was blurred, but the latter grew remarkably in the nineteenth century and was arguably a working-class response to the Victorian 'high culture' of the middle and upper classes. More than anywhere else, in Newcastle music halls provided the main outlet for a distinctive form of popular culture – local songs and ballads. By far the most famous of these is 'Blaydon Races' which has become a Tyneside regional anthem, celebrating the journey from Balmbra's public house and music hall in the Cloth Market out to the Blaydon racecourse. Originally known as the Wheatsheaf (the name Balmbra's coming from its first landlord, John Balmbra), the pub began life – like most of the early 'music halls' – simply as a room in which music was performed. Like many others, it went through a bewildering variety of name changes: by 1859 it was known as the Royal Music Hall; in the 1890s, although not named on the map

OPPOSITE: Charles E. Goad, *Goad Fire Insurance Plan of Cloth Market area* (1896) [NCL]
The Wheat Sheaf is at numbers 6 and 8 and labelled 'P.H. Concert Room & Billiards'. P.H. stands for Public House.

detail shown here, it was the Wheat Sheaf Music Hall at numbers 6 and 8, Cloth Market, then the Oxford Music Hall and subsequently the Carlton Hotel before reverting to Balmbra's in the 1960s.

'Blaydon Races' was written by Geordie Ridley, a Gateshead man who started work in the mines at the age of eight and then became a cart driver. Seriously injured by a waggon that ran out of control, he had to give up manual work, turning instead to performing as a means of support. He sang in many of the music halls and began to write lyrics that celebrated local events and local people, all written and sung in the Geordie vernacular. Ridley died aged only 30 but was one of the prominent local balladeers alongside others such as Blind Willie Purvis (1752–1832), a virtuoso fiddler and great favourite of the Quayside keelmen, Joseph Robson (1808–70) and Joe Wilson (1841–75).

One of Robson's songs was 'The Paanshop's Bleezin', in which women mourned the loss of their pawned clothes and jewellery. Another was political, 'The Horrid War in Sandgate', which – with many 'racial' slurs against the Irish – recounted the riot of 1851 when local Sandgate people turned on the Irish who had moved to the area after the famine. Joe Wilson wrote the hugely popular 'Keep yor feet still Geordie Hinny', focusing on the common experience of two working men, Bob and Geordie Johnson, who economised by sharing a bed in a lodging house. Some of the earlier well-known songs include the eighteenth-century 'Keel Row': 'As I came thro' Sandgate / I heard a lassie sing: / O weel may the keel row / That my laddie's in.' An earlier and more widely known song, 'Bobby Shafto', was used in the eighteenth century to support Robert Shafto, MP for County Durham, by adding a new verse: 'Bobby Shafto's looking out / All his ribbons flew about / All the ladies gave a shout / Hey for Bobby Shafto.'

But it is Ridley's song that is best known. Like most of the writers, he used existing tunes against which to set many of his lyrics. His 'Cushie Butterfield', for example, was sung to the tune of London's 'Pretty Polly Perkins'. It poked fun at the sellers of whitening-stone – the 'yella clay' widely used to clean and decorate the steps of the ubiquitous terraced houses: 'er

nyem's Cushie Butterfield, an' she sells yella clay / An' her cusin is a muckman, an' they caall him Tom Gray. / She's a big lass an' a bonny lass / An' she likes hor beer / An they call her Cushie Butterfield / An aw wish she wor here.'

The choruses of these songs were belted out with gusto in the pubs and music halls of the time – indeed, some still are. But 'Blaydon Races' continues to be by far the region's favourite. Ridley placed it squarely in the Tyneside of his time. The races were originally sited on Blaydon Island and first held in 1861. Most of the islands in the Tyne were cleared by the Tyne Improvement Commissioners, and such was the fate of Blaydon Island in the late 1860s. However, a further site, on Stella Haugh to the west of Blaydon, allowed races to start again in the 1880s, and they continued there until 1916.

The journey to Blaydon featured in the song was by horse-drawn omnibus from Balmbra's Music Hall following a route along Collingwood Street to Scotswood Road, an important turnpike road to the west as shown in Grace's 1828 map. The route crossed the Tyne over the chain bridge to Blaydon and on to Blaydon Island. The places mentioned in the lyrics would have been very familiar to locals: Armstrong's factory was the major heavy engineering firm on the banks of the river (*see* 1852–53); Paradise was a small pit village on the Benwell estate; and the Robin Adair was one of some 50 pubs flanking Scotswood Road. The Infirmary, to which many of the injured went after the accident when the omnibus wheel fell off (verse four), was just to the west of Central Station; and Dr Gibb, a surgeon at the Infirmary, had by the 1860s opened his own private clinic on Westgate Road.

However, Balmbra's has a special place in the history of Newcastle's burgeoning music halls, theatres and palaces of variety. This was recognised with an opportunistic example of place promotion in 1962, when Newcastle Council leader T. Dan Smith organised a major carnival with a great procession to Blaydon. Ostensibly this was designed to celebrate the centenary of the Blaydon Races but was also used to display the newly redeveloped Scotswood Road area, complete with high-rise, system-built flats and modernist sculptures. Smith's hope that this would become an annual festival did not materialise.

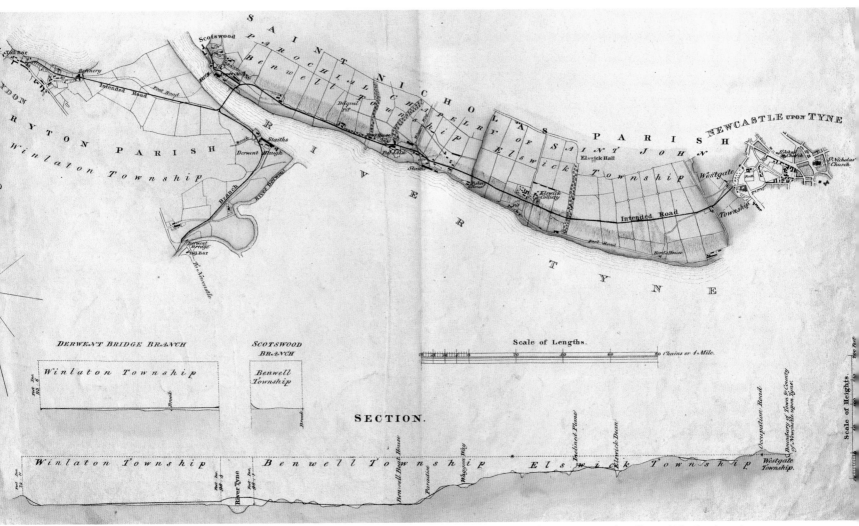

E. Grace, *Plan and Section of the Proposed Turnpike Road from the Town & County of Newcastle upon Tyne to the village of Blaydon* (1828) [NCL] Blaydon horse races were held in at least three different locations. At the time of the writing of the song (1862) they were held at Dents Meadow Island in the River Tyne.

Although Balmbra's Music Hall was revived by the event, this was not to last as it fell into decay along with most of the east side of the Cloth Market. Now, however, there are proposals for regenerating the area that include the restoration and conservation of the building. It was there where Geordie Ridley first sang 'Blaydon Races' and went on to perform in many of the other theatres and halls in the town. He never fully recov-

ered from his injuries but, while he died at a sadly tender age, what a splendid legacy he left the area. Ridley's choruses have lasted for over 150 years: 'Oh me lads, yu shudda seen us gannin / Passin the folk along the road, just as they were standin / Thor wus lots o' lads an' lasses there aall wi smiling faces / Gannin along the Scotswood Road to see the Blaydon Races.'

1899

Shipbuilding on the Tyne

Its banks from Scotswood to the sea resound with the din of the rivetter's hammer. Shipyards are to be seen at every bend on both banks.

R.J. Charleton,
The History of Newcastle upon Tyne, 1885

The Ordnance Survey map of 1899 at the scale of six inches to one mile illustrates well the peak of the shipbuilding industry and its associated infrastructure on Tyneside. The map also illustrates the fascination with and utility of mapping at a relatively large scale. This was a matter of heated debate when the Ordnance Survey was in its infancy resulting in the 'battle of the scales'. Initially, mapping at the scale of the map shown here was favoured. Built-up areas with street names (although not the detail of individual properties) are clearly shown, along with public buildings,

field boundaries and rights of way, infrastructure such as railways, sidings and tram tracks, water features and spot heights above sea level. But especially noteworthy, as illustrated on this map, is the very substantial detail given of industrial and ancillary buildings, usually identified by type of function.

Tyneside has a long and venerable history of shipbuilding dating back over many centuries and providing the region with some of its defining images. All Saints Church, standing above the Quayside, contains an impressive plaque dedicated to Thomas Wrangham (1627–89) who 'built 45 sailing ships'. In fact, Wrangham's seventeenth-century business was unusual in terms of its size and longevity. In those early years it was more common for several shipwrights to form a temporary partnership, rent a suitable piece of riverside ground, build a vessel, sell it and then disband. Two key

OPPOSITE: Ordnance Survey, detail of *River Tyne from Walker to West Jarrow* (1899) [NLS]
In 1899 no fewer than 122 ships were completed on the Tyne.

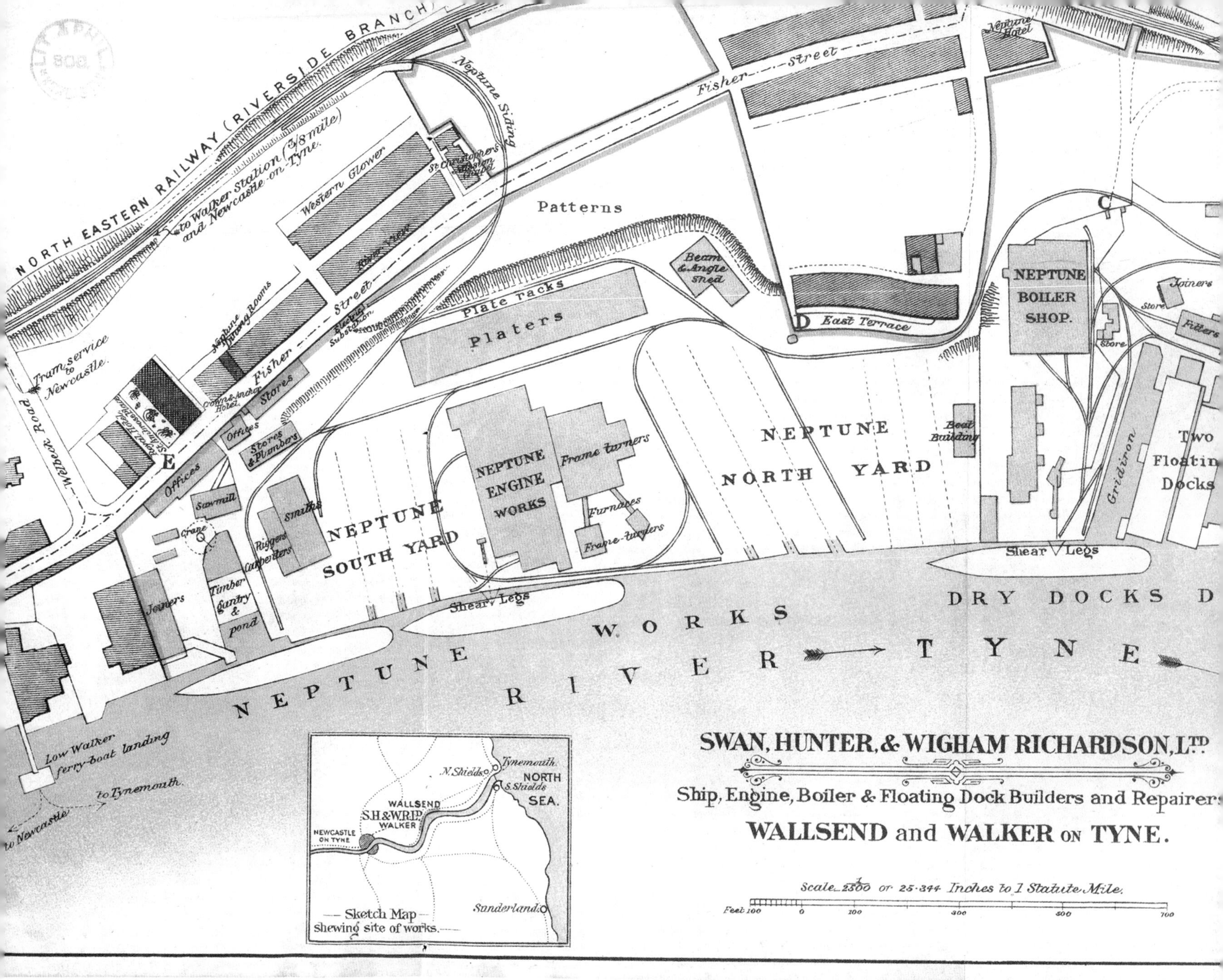
NORTH EASTERN RAILWAY (RIVERSIDE BRANCH)
to Walker Station (3/8 mile) and Newcastle on Tyne.
Fisher Street
Neptune Siding
Neptune Hotel
Western Glower
St Christopher's Mission Chapel
Patterns
Beam & Angle Shed
East Terrace
NEPTUNE BOILER SHOP.
Joiners
Store
Fitters
Plate Racks
Platers
Tram service to Newcastle.
Welbeck Road
Crown & Anchor Hotel
Stores
Offices
Stores & Plumbers
Sawmill
Crane
Smiths
Riggers Carpenters
NEPTUNE SOUTH YARD
NEPTUNE ENGINE WORKS
Frame turners
Furnaces
NEPTUNE NORTH YARD
Boat Building
Gridiron
Two Floatin Docks
Shear Legs
Joiners
Timber Gantry & pond
NEPTUNE WORKS
DRY DOCKS
RIVER TYNE
Low Walker ferry-boat landing
to Tynemouth.
to Newcastle
Tynemouth
N. Shields
NORTH
S. Shields
SEA.
WALLSEND
S.H.&W.R.L.D.
WALKER
NEWCASTLE ON TYNE
Sunderland
Sketch Map shewing site of works.
SWAN, HUNTER, & WIGHAM RICHARDSON, LTD.
Ship, Engine, Boiler & Floating Dock Builders and Repairers
WALLSEND and WALKER ON TYNE.
Scale 1/2500 or 25·344 Inches to 1 Statute Mile.
Feet 100 0 100 300 500 700

Swan, Hunter & Wigham Richardson's Shipyards and Works (1905) [LIT & PHIL] Concerned at growing German shipbuilding capacity, the British Government provided an annual subsidy of £150,000 and a low-interest loan of over £2 million to help build the *Mauretania*.

factors were then operating against any large-scale development of the industry – the difficulty of navigation on the river and Newcastle's monopoly of economic activity along the Tyne – but neither could prevent the expansion of the industry once wood became scarce and was gradually replaced as a fuel by coal. Tyneside's exploitation of its coal required a large shipping fleet for exporting its output, so

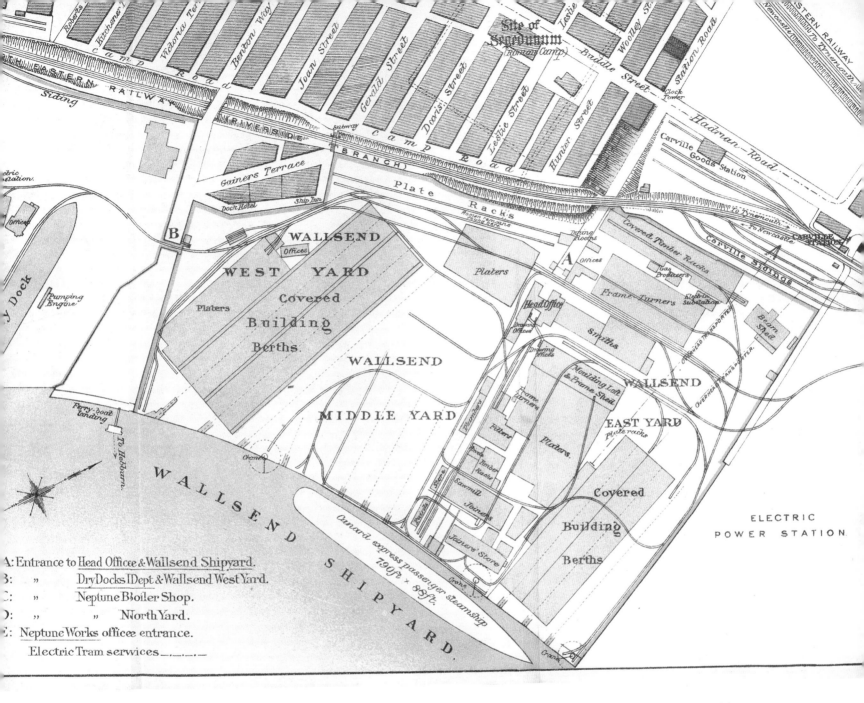

A: Entrance to Head Office & Wallsend Shipyard.
B: "   Dry Docks I Dept & Wallsend West Yard.
C: "   Neptune Boiler Shop.
D: "   "   " North Yard.
E: Neptune Works office entrance.
   Electric Tram services ___.___.___.

large-scale shipbuilding on the Tyne was a product of the demand for coal.

The principal output of the earlier yards was of relatively small colliers. In the later eighteenth and early nineteenth centuries, around 20 shipyards were operating on Tyneside, the main locations then being at Newcastle and South Shields, although two larger builders – Francis Hurry and Edward Mosley – developed at Howdon, each building over 50 ships during the later eighteenth century. The American War of Independence and, especially, war with France led to the expansion of the Royal Navy, and growth in demand for larger fighting ships, which was met on the Tyne by firms such as the Temple's, initially at South Shields but subsequently at Jarrow on a site later developed by Charles Palmer.

By this stage, there was already some innovation. Examples include the invention of the self-righting lifeboat by Henry Greathead and William Woodhave of South Shields in 1802, the building of the Tyne's first steamboat in 1814 and the inauguration of a faster river-ferry system, with the *Tyne Steam Packet* (later the *Perseverance*) taking passengers between Newcastle and the Shields. However, from around the middle of the nineteenth century, accelerating innovation in marine engineering placed the region at the global forefront of the industry. In 1842 the *Prince Albert*, built by Coutts at Walker, was one of the first iron-built steamers in the world, while even more significant was the launch of a screw-propelled iron collier, the *John Bowes*, in 1852 by the new firm of Palmers at Jarrow. The *John Bowes* took five days to complete the round trip from the Tyne to London, a voyage that could take sailing ships up to a month. The 'golden age' of Tyne shipbuilding had begun, and Palmers was to play a central part in its heyday. The 1854 Crimean War increased the demand for iron-plated ships, and Palmers was again innovative, building the 2,000-ton battleship the *Terror* in only three months. New (and subsequently famous) firms emerged, such as Charles Mitchell at Low Walker, Andrew Leslie's at Hebburn and John Wigham Richardson taking over the former Coutts yard at Walker.

By the later nineteenth century North East England's shipbuilding (stretching 50 miles from Middlesbrough to Blyth) dominated the world industry, producing over 40 per cent of global output in some years. Tyne firms had diversified and, in addition to cargo vessels, were producing ocean liners and large warships. Typical of this innovation, Palmers built early oil tankers and developed new techniques such as rolling instead of forging armour plates. Innovation continued with steel replacing iron and hydraulic riveters replacing hand riveting.

If the early years of shipbuilding on the Tyne were characterised by rapid turnover of small firms, the later nineteenth and early twentieth centuries featured a series of takeovers and mergers, out of which grew a limited number of giant firms. An example is shown by the map of Swan, Hunter &

Wigham Richardson, a firm that came into being in 1903 from the union of three adjoining firms, all of which had grown from small beginnings. John Wigham Richardson's Neptune Yard at Low Walker on the west of the site had only three berths in 1860, employed around 200 men and had a river frontage of around 300 feet. From 1879 an excellent reputation was won by its marine engineering and boiler manufacture, still apparent on the 1905 map. George Burton Hunter joined the nearby firm of C.S. Swan & Co. in 1880. Although larger than the Neptune Yard, its employment was only 717 at that time, with the works spread over about seven acres and a river frontage of only 270 feet.

In its first ten years the new company of Swan, Hunter & Wigham Richardson launched an average of 93,000 gross tons per year from their 17 slipways. The map shows an unbroken river frontage of 4,000 feet and works covering 80 acres, plus floating docks, a large dry dock and huge covered berths for all-weather working. The firm could deal with multiple orders and all aspects of shipbuilding and repairing. They launched probably the Tyne's most celebrated vessel, the record-breaking *Mauretania*, in 1907, when the firm was the second largest shipbuilder in the world. In addition to ocean-going liners, its building of warships, especially during the First World War, was outstanding: 55 warships were launched with an aggregate tonnage of 100,000 gross, with over 290,000 gross tons of merchant shipping also being built at this time. The firm exceeded this record between 1939 and 1945 when it launched 125 ships with an aggregate gross tonnage of nearly half a million. Wallsend, like Jarrow, became essentially a one-industry company town.

This section of the river was virtually one continuous shipbuilding zone in the early twentieth century. The maps of the Armstrong-Whitworth yards tell a similar story. C.W. Mitchell, from Aberdeen, established himself as a successful shipbuilder at the Low Walker yard in 1852. Mitchell built up a strong relationship with Russia, a significant incentive for W.G. Armstrong to seek a partnership with his firm in 1882. Armoured vessels were built at Elswick and other types of vessels at Low Walker. This arrangement flourished, and

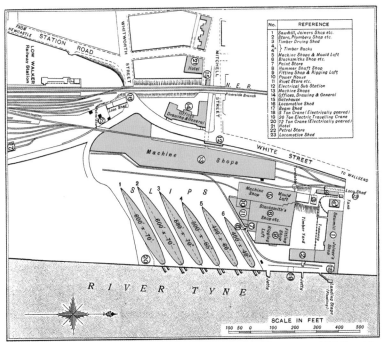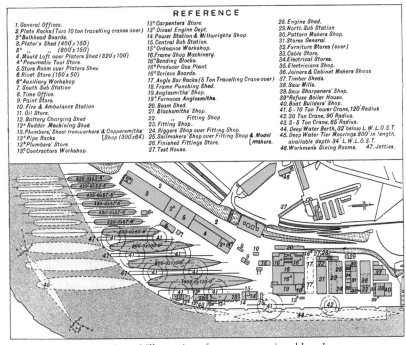

Armstrong-Whitworth, *Low Walker Yard* (LEFT) and *Walker Naval Yard* (RIGHT) (*c.*1950) [SANT] Illustrations from a promotional brochure produced after the Second World War.

in 1897 Armstrong-Mitchell's amalgamated with Whitworth's, a Manchester armaments firm, and its Low Walker yard launched one of the most innovative and famous vessels to be built on the Tyne, the SS *Baikal*, an ice-breaker and rail ferry ordered by Russia. After the launch it was dismantled into 6,900 pieces and delivered to Lake Baikal in 1896 via the Trans-Siberian railway. The ever-increasing demand for larger vessels prompted the firm to develop a second yard just upstream of Low Walker. Built around 1910 and known as the Walker Naval Yard, this occupied a 70 acre site and was equipped with nine berths and ten-ton cranes. In effect, it replaced the Elswick shipyard. Extremely active during the First World War, over 3,500 men were employed here in 1915.

Towards the end of the nineteenth century, yet another major innovation had taken place. Charles Parsons invented the steam turbine engine in 1884 whilst working at the Gateshead firm of Clarke Chapman. He set up his own firm,

the Marine Steam Turbine Co. Ltd, at Heaton, where he built *Turbinia* (still proudly on display at Newcastle's Discovery Museum), capable of achieving the unheard-of speed of over 34 knots. A sceptical Admiralty was astonished at his uninvited demonstration at the Naval Review at Spithead in June 1897. Parsons established premises at Wallsend to build turbine engines, although the ships themselves were built by Hawthorn Leslie and Palmers.

The closure of Palmers shipyard in 1933 was a clear indication that the Tyne's great days of shipbuilding were over. But the region could be proud of its enormous contribution to the industry: spearheading the transition from wood to metal ships; playing a significant role in the invention of the self-righting lifeboat; applying steam power; demonstrating the advantages of water ballast instead of solid ballast; developing oil tankers; manufacturing naval armaments; and inventing the steam turbine as a means of marine propulsion.

1904

Elswick-Benwell: the making of a working-class suburb

Elswick, where an army of workmen toil night and day, fashioning mighty weapons of warfare and no less mighty engines for the service of commerce and peace.

R.J. Charleton, *The History of Newcastle upon Tyne*, 1885

The popular image of the west end of Newcastle is presented in this lead map, that of a terraced nineteenth-century industrial and predominantly working-class residential area. Industrial sites and infrastructure spread westwards along the north bank of the Tyne, and responding to the workers' need for housing, tightly packed terraced housing was built from the riverside up the steep riverbanks. However, the processes that produced this townscape are somewhat more complex than this simple description implies, with the relationship between industrial development and housing provision varying in terms of timing and type.

Historically, this area was largely the jurisdiction of the manor of Elswick and its neighbouring manor to the west, Benwell. By the late eighteenth century, these had come into the ownership of the Hodgson and Shafto families respectively and were predominantly agricultural estates, though increasingly being exploited for coal. Over time these large estates were divided into smaller (although still quite substantial) plots on which a number of mansions were built. By the early nineteenth century, the 700-acre Elswick estate was in the ownership of the Newcastle MP John Hodgson, an early opponent of allowing the Newcastle–Carlisle railway to cross through his land. Conscious of the advancing tide of industrial urbanisation, Hodgson sold out to Richard Grainger and moved to Hexham.

Grainger saw the development of the west end as a mixed residential and industrial area and as the logical next step to

---

OPPOSITE: Andrew Reid, detail of *Reid's Map of Newcastle and Gateshead* (1904) [TWA] Purdue's description of 'phalanxes of Tyneside flats marched with parade-ground precision up the banks' is highly appropriate.

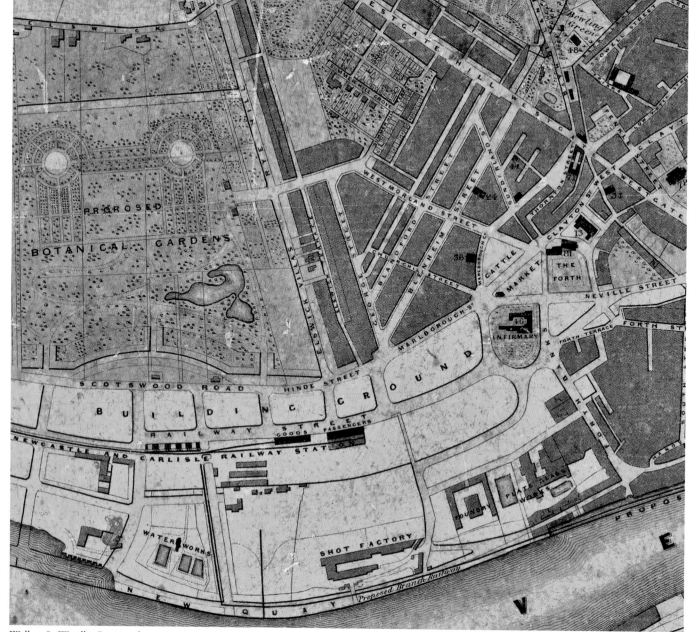

Walker & Wardle, *Proposed Botanical Gardens* (1844) [GLH] Letters from John Hancock (1808–90) refer to a meeting at the Lit & Phil in February 1839 to discuss the establishment of botanical and zoological gardens on land to be leased from Mr Grainger.

his work in central Newcastle. He moved to live in the palatial Elswick Hall. Although no overall physical plan survives, it is known that he started by building middle-class housing in the Rye Hill area and the Grade II listed terrace Graingerville West. But his plans were much more ambitious and included the development of botanical gardens with a zoo, with the support of the Natural History Society. In terms of attracting industry to the area, he proposed to build a railway terminus near the river at the eastern end of the estate, this being prior to the building of Dobson's Central Station. However, Grainger had over-mortgaged himself and, advised by the town clerk, John Clayton, who also acted as his solicitor and financial advisor, he was forced to satisfy his creditors (mainly the Northumberland and Durham District Bank) by selling off parcels of his land piecemeal. The density and type of housing that this process produced bore very little relationship to the trajectory of development of local employment. George Cruddas, one of Armstrong's partners and his chief financial advisor, had bought 48 acres of the western part of the Grainger estate in 1861–62, but housing development did not take place until much later. The process continued long after Grainger's death, his affairs only being finally settled in 1901. Meanwhile, somewhat ironically, the Northumberland and Durham District Bank was itself in liquidation and this increased the pressure to develop remaining parts of the estate.

The development of the Benwell estate came somewhat later. Its southern part was bought by the famous coal viewer John Buddle in 1811. His interest was primarily in the coal resources of the Delaval, Paradise and Charlotte pits. Buddle died in 1843, and the estate passed to his nephew and was managed by Sir William Armstrong and Benjamin Chapman Browne as trustees. Despite the fact that Armstrong had started his works in 1847 immediately adjacent to this, it was not until the 1880s that significant housing development took place. The plans for a 'new town' in South Benwell, announced in 1883, featured a classic late Victorian/Edwardian working-class terraced area, with subtle variations in social status reflected in the details of housing design which, nonetheless, promoted a private, family-oriented lifestyle. The most basic, as shown in Northbourne Street, was the 'Tyneside Flat' of two or three rooms although the upper flat shown here had the luxury of a further attic bedroom. These terraced flats were unique to the region, with the adjacent front doors of the upstairs and downstairs flats opening straight onto the pavement. An important component was the private back yard, containing toilets and coal storage. A slightly more desirable variation was where the pair of flats had a bay window. Next up the scale were the houses aimed at supervisory workers, foremen and lower-grade clerical workers, these comprising two storeys, with their higher status reflected in a bay window and small front garden. At the top of the social gradation were houses with a greater degree of exclusivity, signalled by their streets being pedestrian rather than open to general traffic and with their houses sited further from the river and Armstrong's vast works.

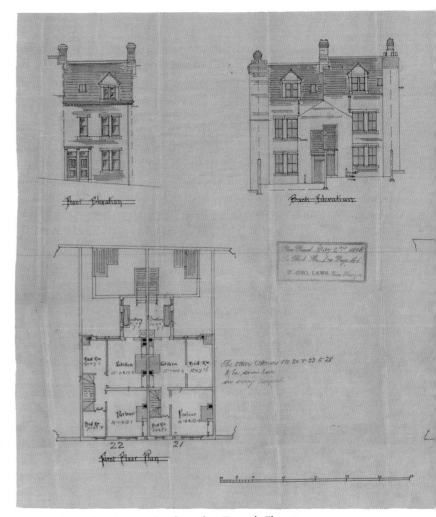

Front and rear elevation and first floor plan, Tyneside Flats at Northbourne Street, Elswick (1888) [TWA] The 1911 census showed that 55.7 per cent of Newcastle's population lived in flats.

Much of the terraced housing in west Newcastle therefore arrived years later than the expansion of the employment there. One important factor in this disjointed timing was that by the 1880s working-class real wages were increasing, particularly for skilled workers, partly as a result of increased pressure from organised labour. This made the rental income gained from working-class housing development a much more attractive proposition for both builders and landlords.

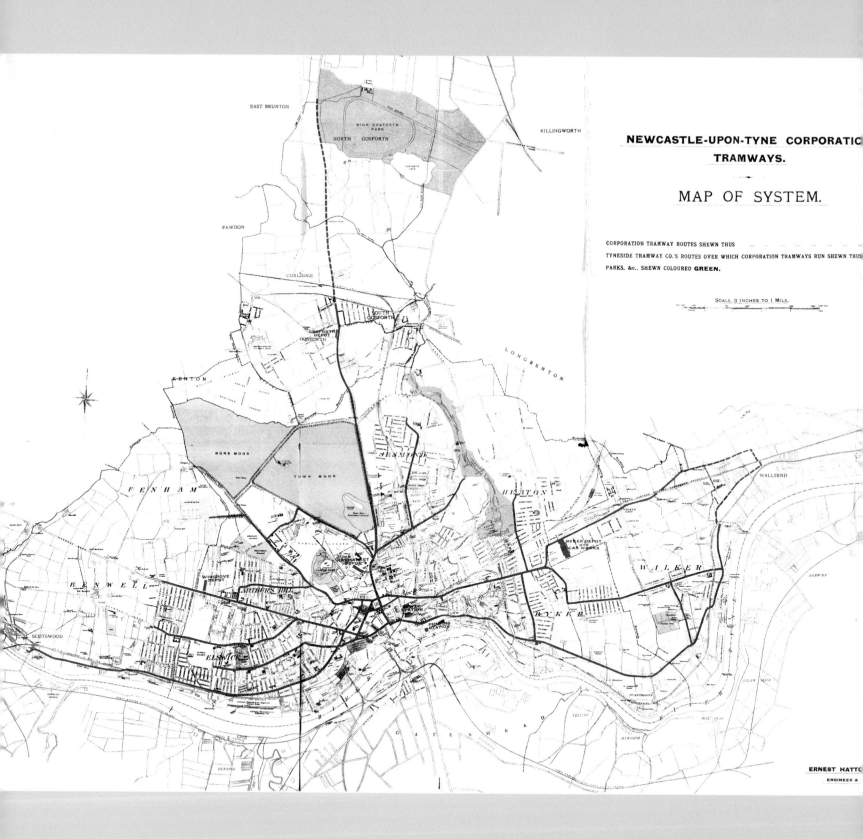

NEWCASTLE-UPON-TYNE CORPORATION TRAMWAYS.

MAP OF SYSTEM.

CORPORATION TRAMWAY ROUTES SHEWN THUS

TYNESIDE TRAMWAY CO.'S ROUTES OVER WHICH CORPORATION TRAMWAYS RUN SHEWN THUS

PARKS, &c., SHEWN COLOURED **GREEN**.

SCALE, 3 INCHES TO 1 MILE.

ERNEST HATTO

ENGINEER &

1908

Public transport: the growth of a tramway system

Previously a worker either must live very close to his work or must face long walks in all kinds of weather. With the coming of the tram and the extension of suburban trains . . . it was increasingly possible for workers to live at some distance from their place of work.

N. McCord, *North East England*, 1979

The main map shows that Newcastle had a comprehensive tram system by 1908, but its evolution was hardly simple or straightforward. In the second half of the nineteenth century, railway development provided an effective transport system both across the northern region and linking with other regions. But within Newcastle itself, transport provision was extremely limited, and the idea of providing it for the mass of the population aroused little enthusiasm. The majority of workers walked to work, as a traffic survey on the two bridges over the Tyne

in 1859 showed. Over six days, 168,098 people walked over the 'old' Tyne Bridge and 27,268 over the High Level Bridge. A proposal from private enterprise to develop a 'horse-drawn street railway' in 1862 was met with apathy at best and, when the idea was resuscitated in 1871, Councillor Gregson described the scheme as 'a source not only of great inconvenience but a very great danger'. However, by 1876 a report on tramway development to the Town Improvement Committee was favourably received, and the full Council voted in favour of developing a scheme which would be leased to private operators. A significant factor in this change was the gradual realisation that Newcastle was falling behind other urban areas in its provision of public transport, and this created a more receptive environment for those arguing that public transport could actually increase industrial efficiency and deal with some of the problems of rapid urban growth.

OPPOSITE: Andrew Reid, *Newcastle upon Tyne Corporation Tramways: Map of System* (1908) [NCL] The Corporation network was concerned solely with provision internal to the city. External links to the north and east were provided by the Tyneside Tramway Co.

Detail from *Bird's-Eye View: Royal Mining, Engineering and Industrial Exhibition: Horse-drawn trams on Great North Road* (1887) [TWM] Horse-drawn tramways operated in Newcastle from 1878 to 1901.

Even so, the initial horse-drawn tram system in Newcastle provided an extremely basic service. Though there were routes from Central Station to Osborne Avenue in Jesmond, to the Minories at Sandyford and to Byker, and also routes from Grey's Monument to Scotswood, Elswick Road and Gosforth, they totalled only eight miles (mostly just single-track) and were operated by only 44 tramcars and 272 horses, even by 1898. By then, however, growing complaints about the lack of comfort and reliability of the service led to demands for its municipalisation. Unusually for Newcastle at this time, with a Council dominated by private business interests, the complaints proved persuasive, and in 1901 the Council took control of the tramway service and rapidly decided that it should be electrified. The first three such routes were from Central Station and went to Osborne Road in Jesmond (ending at its junction with St George's Terrace), to Brighton Grove in Westgate ward and along the Great North Road to the borough's northern boundary with Gosforth. The impact of electrification was almost immediate: whereas the horse-drawn tram system had carried 6,486,362 passengers in 1898, in 1903–04 the electrified system carried 39,715,120.

The quotation from McCord is a little ingenuous as initially the electrified system favoured the middle-class areas of the city. Of the first six routes electrified, only two served working-class districts, and there was also a much lower frequency of service for the latter. For example, the working-class district of Elswick Road had a service of one tramcar every 20 minutes compared to every few minutes on Osborne Road in Jesmond. Nevertheless, the electrified system expanded rapidly, with 21 miles of track in 1904 (compared to 15 in 1901) and with six further extensions up to 1913. Moreover, as the lead map indicates, much of this additional mileage now linked working-class residential districts to industrial areas: the most heavily used route was along Scotswood Road, with 11 million passenger journeys, over half of which were accounted for by tramcars for 'workmen', and with over 100 special tramcars running before 7 a.m. with reduced fares.

However, by the early twentieth century, Newcastle was but one part of a growing conurbation, consisting of several individual administrative jurisdictions. A major problem therefore concerned the provision of services – especially transport – to unify these areas. Tramways were most usually provided separately by each local authority, leading to lack of co-ordination across the conurbation, not least in the use of different gauges of tram line by neighbouring tramway systems such as Newcastle–Tynemouth and South Shields–Gateshead. The Jarrow and District Electric Tramway Company started their service to Tyne Dock in 1906 but, again, use of a different gauge prevented through-running to South Shields. The Tyneside Tramways and Transport Company, founded in 1901, represented a rare attempt to provide cross-boundary services as it developed lines from Wallsend to Gosforth and North Shields and, in 1904, from Gosforth to Gosforth Park.

There had been an early attempt to link Gateshead with Newcastle, with the introduction in the 1880s of the 'ha'penny lop' (a single coach pulled by one horse and carrying up to 40 people) across the High Level Bridge. The main constraint on expanding the services here was the toll charged by the North Eastern Railway who owned the bridge. The resolution to this obstacle did not occur until 1920 when Newcastle Corporation obtained permission to build a double line across it. The tramway opened in 1923 with a service journeying from the Cloth Market in Newcastle to the War Memorial at Shipcote,

Gateshead. Shortly after this, services were being provided from Newcastle to Dunston, Bensham, Low Fell and Heworth and, with the opening of the new Tyne Bridge in 1928, a further link was provided.

Gateshead's first tramway network was, intriguingly, a steam-powered one (an experiment with this in Newcastle had been a failure), this operating on three routes – eastwards to Heworth along Sunderland Road, westwards to Teams and south to Low Fell. In 1897 the system was taken over by British Electric Traction Ltd and converted to electric traction. Existing tracks were doubled and some lines extended. For 20 years the Gateshead tram system flourished and was even the subject of an admiring review by the *Newcastle Chronicle* in 1905. However, rapid deterioration set in during and after the First World War, but the threat of an official enquiry led to significant improvements. In 1921 the Company carried over 26 million passengers and, boosted by these numbers and the confidence of knowing they had sanction to operate until 1936, the Company modernised the system with double tracks and new cars, producing a large-scale network, as the 1937 map shows. However, it is particularly noticeable that riverside industrial areas had no service. Trams were not primarily intended for industrial workers. The 1937 map was actually produced with the intention of introducing trolley buses on some existing tram routes. Although, in theory, trolley buses were more flexible than trams in terms of the routes they could take, there was no obvious attempt to intensify route provision. Rather, the proposals were to extend existing tram routes, for example in the east to Bill Quay and Wardley Square, route number 6 to Lobley Hill in the south-west and route number 4 to the west of Team Valley. Although this reflected patterns

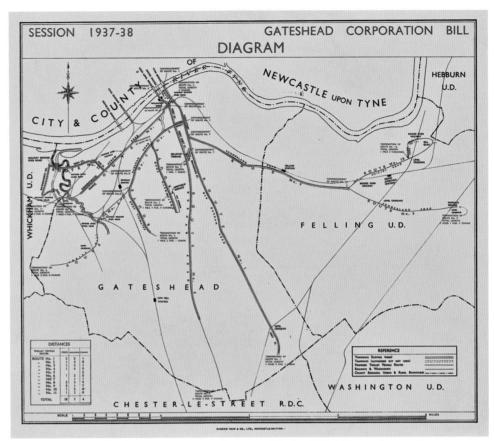

Andrew Reid, *Gateshead Tramway system and proposed Trolley Vehicle Routes* (1937–38) [GLH] The proposed trolley vehicle routes were clearly intended to serve suburban areas which had grown substantially in the interwar period.

of residential growth to some extent, it still would have left large areas, particularly in the eastern part of Gateshead, with limited access to public transport.

In the 1930s, trams had become increasingly subject to criticism, not least from owners of motor vehicles competing for road space. Newcastle Corporation introduced trolley buses in 1935 and Gateshead Corporation was clearly attempting to do so from 1936. However, the latter's plans were thwarted by the Tramways Company, which eventually morphed into Gateshead District Omnibus Company, running motor omnibuses to Wrekenton and Heworth. Gradually, trams were phased out. In Newcastle, however, trolley buses provided an intervening mode between trams and omnibuses, at least up to their demise in 1966. At their peak there were 28 trolley bus routes in the city and a fleet of 204 trolley buses.

1924

The Newcastle upon Tyne Society: towards a planned future?

The latest critic is a popular novelist, who while admitting 'a certain sombre dignity' in central Newcastle is scathing in his references to the neighbourhood generally.

The Newcastle upon Tyne Society,
Annual Report, 1933

The 'popular novelist' was, of course, J.B. Priestley whose somewhat unselfconscious criticism (he was from Bradford) sparked the resentment of many within the region. But, to its credit, the Newcastle upon Tyne Society, formed in 1924 'to promote a wider concern for the beauty, historical interest, amenity, healthfulness and development of the city', chose to interpret it differently. Its view was that such criticism was a necessary shield against complacency and a possible catalyst for improvements. Like so many English voluntary institutions,

the Newcastle upon Tyne Society was conceived over a luncheon, in this case of the Newcastle Rotarians. The broad base of the organisation was signalled by the early association of two rather unlikely individuals, Colonel J.D. Mitchell and Councillor David Adams, the former a fervent advocate of architectural aesthetics and the other a fierce proponent of improved housing and social conditions for the working classes. However, as might be expected, the membership was overwhelmingly of middle-class professionals like architects, lawyers, academics and antiquarians – but, crucially, it also included individuals with a commitment to the emerging field of town planning. Foremost amongst these was the architect Robert Burns Dick, a partner in the architectural practice of Cackett, Burns Dick & Mackellar.

In 1924, Burns Dick designed the plan for the city centre shown here. In several respects, his plan had a strong influence

DETAIL OPPOSITE AND OVERLEAF: Robert Burns Dick, *Burns Dick Plan for Central Newcastle* (1924) [NCL] The influential architect and planner H.V. Lanchester lavished high praise on the plan, describing it as 'a fine piece of imaginative work'.

suggested by J.T. Cackett in 1905. Much concerned with catering for motor traffic but diverting through journeys away from the city centre, he also proposed a city-centre bypass around its western edge and a new road leading north-west directly from the Redheugh Bridge, as well as significant extensions to Northumberland Road, both eastward to give access to Heaton and beyond, and westward across the northern tip of Northumberland Street to the western edge of Gallowgate and leading on to the top of Westgate Road.

But the most eye-catching feature was a new Civic Centre on the south-east corner of the Town Moor. The approach from the south would be by way of a wide boulevard leading from Barras Bridge and passing to the west of the Hancock Museum to a monumental Civic Centre, set in a complex of municipal buildings. The model for this proposal was Cardiff's Civic Centre and, rather more grandiosely, Burns Dick noted that 'Such a site is that of the Capitol at Washington'. Although the final choice of site for a new Civic Centre (*see* 1890) was much delayed, the Burns Dick plan was significant in promoting a site to the north of the centre at a time when the balance of opinion was swinging in favour of a 'make do and mend' approach to renovating the constricted Town Hall site in St Nicholas Square.

It is clear that the Newcastle upon Tyne Society concerned itself with a broad range of planning and environmental issues and, reflecting national concerns in the interwar period, one of these was the idea of a Green Belt. As elsewhere on Tyneside, suburban growth was accelerating, including speculative land subdivision into 'plotlands' such as on the Darras Hall estate. Responding to such concerns, the Society produced a plan for 'The City Green Belt', finalised in January 1927 and submitted to Newcastle City Council, which had already begun to consider a development plan for the city. The Society's proposal took a rather different form from the wide belt of countryside that Ebenezer Howard, in his 1898 book *Tomorrow*, had put forward for his 'garden city' and then became one of the central features of post-war land-use planning on Tyneside, as around other large cities. Instead, it incorporated the approach of 'green girdles' and parkways being developed in North America

on subsequent developments, if not in terms of precise locations then in terms of ideas. Recognising the growing importance of motor transport and the imminent construction of the Tyne Bridge, the plan featured several significant new roads. A major intersection was proposed immediately to the north of the new bridge (a little south of the current Pilgrim Street roundabout), with new roads leading west to Central Station, one leading south-east to access the Quayside and one running east of Northumberland Street, virtually on the line of the present John Dobson Street, although the latter had already been

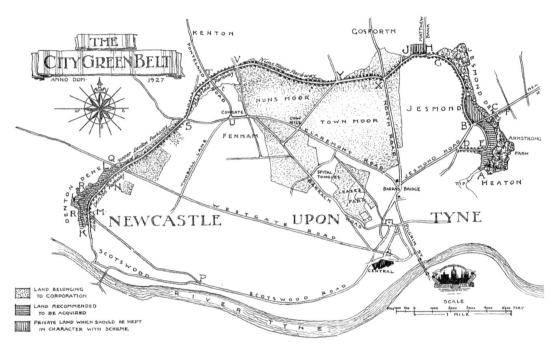

THE CITY GREEN BELT
·ANNO·DOM· 1927

NEWCASTLE UPON TYNE

LAND BELONGING
TO CORPORATION
LAND RECOMMENDED
TO BE ACQUIRED
PRIVATE LAND WHICH SHOULD BE KEPT
IN CHARACTER WITH SCHEME

Robert Burns Dick,
*The City Green Belt:
Plan illustrating the
proposals of the
Newcastle Society*
(1927) [NCL]
Although the map is
headed 'Green Belt',
the use of 'Parkway'
in the western areas
indicates familiarity
with North American
ideas on maintaining
green space in the city.

for recreational and amenity purposes and proposed for London in the early 1900s by the likes of architect-planners William Bull, George Pepler and Raymond Unwin.

As the map shows, the Green Belt proposals suggested that Denton Dene would be developed into a public park, and on the east of the City, Ouseburn Vale down to Heaton – 'naturally very beautiful but neglected and dishevelled' – would 'be saved from further damage and made a Public Park'. A new 'parkway' would link these two parks running from Denton Dene to Nuns Moor and skirting the northern edge of the Town Moor where it would link up with Jesmond Dene Road and run through to Armstrong Park and Heaton. The result would be 'an unbroken belt of park, moor and parkways from Scotswood to Heaton and a good bypass road between the North and West [of the city]'. Successive annual reports of the Society, however, noted little progress, despite the close links between the Society and the City Council. By 1937 the Society's annual report recognised that it was too late for the complete green belt proposed, though a partial scheme was still possible, but then war intervened.

The Newcastle upon Tyne Society remains an important force within planning and environmental issues in the city and region. Indeed, as early as 1929 its wider sphere of geographical interest was apparent when it became the Northumberland and Newcastle Society. With the emergence of town planning as a statutory local government activity, however, planning issues became much more prominent on the local agenda and a much greater source of contention.

At the end of the Second World War, partly prompted by bomb damage, planners began to assert themselves on a national scale, producing visionary schemes that would have destroyed most of the townscape that they inherited. Newcastle's 1945 plan was typical, despite the relatively limited war damage. Drawn up by Percy Parr, the City Engineer, it proposed an almost complete revamping of the town which would have swept away most of the Victorian and earlier buildings. However, many of the ideas contained in this plan were later to resurface in the 1960s. These included encircling the central area by motorways (although Parr called them 'ring routes'), creating an educational precinct to the east of the central area, locating a new Town Hall in the north of the central core, redeveloping Eldon Square and, in the social sphere, forming 'community areas'. But post-war austerity meant that these plans were delayed or abandoned.

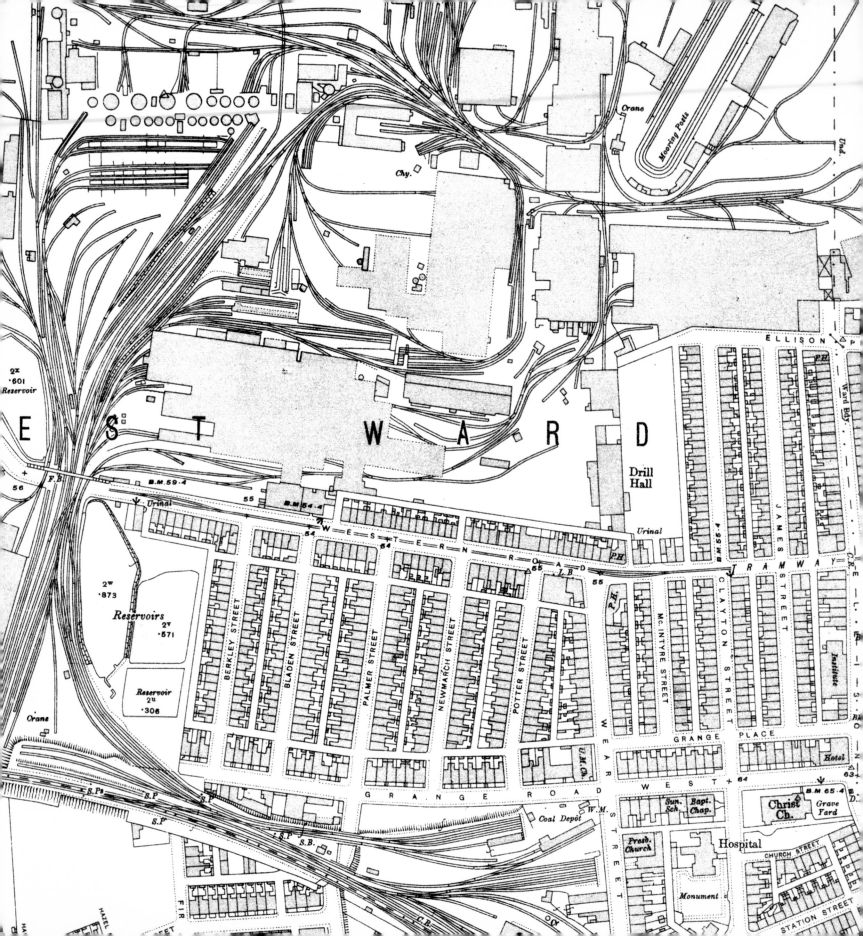

Crane

Mooring Posts

Chy.

EAST WARD

2ˣ
·601
Reservoir

B.M.59·4

56   F.B.

Urinal

BM.54·4

55

Drill Hall

Urinal

ELLISON P

Ward Bdy

P.H.

B.M.55·4

JAMES STREET

C.R.E. L. ... BAR

2ʷ
·873

2ᵛ
·571

Reservoirs

Reservoir
2ᵘ
·306

BERKLEY STREET

BLADEN STREET

PALMER STREET

NEWMARCH STREET

POTTER STREET

WESTERN ROAD

P.H.

L.B.

△·55

55

P.H.

Mc.INTYRE STREET

CLAYTON STREET

TRAMWAY

Institute

Crane

S.Ps
S.P.

S.P

S.P. S.B.

S.P.

GRANGE ROAD

WEAR STREET

GRANGE PLACE

Hotel

GRANGE ROAD WEST

JAMES STREET

·64

Coal Depôt

W.M.

Presb. Church

Sun. Sch.

Bapt. Chap.

Christ Ch.

Grave Yard

B.M.65·4

Hospital

CHURCH STREET

Monument

HAZEL STREET

FIR STREET

G.B.

STATION STREET

1936a

Jarrow: 'the town that was murdered'

Sir Walter Runciman, President of the Board of Trade, said 'Jarrow must work out its own salvation.'

Ellen Wilkinson,
The Town that was Murdered, 1939

The Jarrow March (or 'Crusade' as it was initially called) has entered the halls of fame, not only in the North East but for the whole country. The route of the march has a symbolic significance and has been taken, in part or completely, by numerous protest marchers subsequently. At the time, the men who marched felt that they had been cheated and their efforts a waste of time. History tells a rather different story because, even though nothing changed in the short term, their efforts did help to create a new awareness of the plight faced by areas of high unemployment. After the war, new legislation which created urban and regional planning and the welfare state reflected not only the aftermath of the devastation of war but also offered some recognition of the sacrifices that had been symbolised by the men of Jarrow.

The seeds of discontent that prompted the March were sown early in the interwar period. The economic upturn after the First World War proved only a short-lived respite for North East industry. During most of the 1920s the economy was at best stagnant, and the Great Depression began after the stock market crash on America's Wall Street in 1929. It proved the worst economic downturn in the history of the industrialised world and lasted for a decade. Areas like the North East, which relied primarily on heavy industry and exports, were hit especially hard.

In his social survey of Tyneside in 1926, Henry Mess observed that 'Of all the Tyneside towns, Jarrow is the most homogeneous in character. It is a shipyard town, and very little

OPPOSITE: Ordnance Survey, *Palmers Works and adjacent housing, Jarrow* (1917) [NLS] 25-inch edition of 1917, Durham Sheet NE.

else. It is almost entirely a working-class town.' The map demonstrates the physical dominance of one employer and the associated tightly packed rows of working-class terraced housing whose households depended disproportionately on Palmers shipyard. Before the crash, Palmers itself employed over 10,000 workers, and local ancillary trades and services were critically dependent on it. It had been established in 1852 by Charles Palmer and by 1900 Palmers Shipbuilding and Iron Company manufactured its own steel as well as making ships. Its Jarrow site covered some 100 acres along the south bank of the Tyne. Initially, it had made collier ships, but over time it increasingly attracted contracts for military vessels, supplying navies throughout the world. Its launchings included battle-ships, destroyers, cruisers and river gunboats, as well as civilian cargo ships and tankers. However, during the 1920s most of the customer countries developed their own shipbuilding facil-ities and, with the Depression, by 1930 Palmers' order book was empty and its finances in substantial deficit. Closure was briefly averted by an Admiralty contract but, by 1933, the National Shipbuilders Security (formed in 1930 to acquire and dismantle failing yards) bought Palmers and began to dismantle the shipyard, engine works, blast furnaces and rolling mills.

The impact on Jarrow was catastrophic, with levels of male unemployment reaching over 70 per cent. The town had a brief respite from a most unlikely source, Sir John Jarvis, High Sheriff of Surrey who became a Conservative MP for Guildford in 1935. He launched an appeal, the 'Surrey Fund', which raised an impressive £40,000 which he used to create 'make-work' jobs. As recorded on a commemorative plaque by the Borough of Jarrow, the funds were used, among other things, 'to brighten the homes of the people of Jarrow (and) to construct the Monkton Dene Park'. He also used his personal wealth to buy two large, decommissioned liners which were brought to the Tyne to be broken up by a new shipbreaking company based at Palmers yard, with the retrieved metal then used in the manufacturing company that he founded in the area, Jarrow Metal Industries Ltd. He was made a Freeman of Jarrow in 1935, although the ceremony was boycotted because of his becoming a Conservative MP just before. Nor were his

efforts universally applauded since many were suspicious of philanthropy and saw charitable donations as no substitute for national action to tackle unemployment and poverty.

The March in 1936 aimed to draw Parliament's attention to the plight of the town. The idea was enthusiastically supported by the Council, the town's MP Ellen Wilkinson and a range of religious and political groups. Some 200 unemployed men, carefully chosen to ensure that they were physically fit, were selected to take a petition to present to Parliament which read, rather blandly:

During the last 15 years Jarrow has passed through a period of industrial depression without parallel in the town's history. Its shipyard is closed. Its steelworks have been denied the right to reopen. Where formerly 8,000 people, many of them skilled workers, were employed, only 100 men are now employed on a temporary scheme. The town cannot be left derelict, and therefore your Petitioners humbly pray that His Majesty's Government and this honourable House should realise the urgent need that work should be provided for the town without further delay.

The marchers started out on the route shown on the accom-panying poster on 5 October and arrived at Marble Arch on 31 October to coincide with the opening of Parliament. It was a journey of almost 300 miles. The poster shows the feisty Ellen Wilkinson, who had joined the March but left at Chester-le-Street to attend the Labour Party's annual conference in Edinburgh, where she had a frosty response and a refusal to support the March. She later rejoined the March and presented the petition to Parliament on 4 November. There was, however, barely any discussion of the petition. Moreover, Prime Minister Baldwin had refused to meet the marchers, while the President of the Board of Trade Walter Runciman, whose father was a major North East shipowner, bluntly told the Commons that shipbuilding could not be revived in Jarrow and that the town's unemployment situation had shown some recent improvement. The marchers themselves discovered that the Unemployment

Assistance Board had reduced their benefits because they were unavailable for work while away. The fear of civil unrest was clearly paramount in the minds of politicians. Indeed, perversely, it could be argued that Jarrow was unfortunate that it was Ellen Wilkinson who gave the town political leadership. She served as its Labour MP from 1935 to her early death in 1947, by which time she had risen to be Minister of Education. But in her earlier life she was an Independent Labour Party member with Marxist leanings and, in 1920, became one of the founding members of the Communist Party of Great Britain, even though she also kept her Labour Party membership. She would undoubtedly have been seen by most politicians as a firebrand to be treated with caution.

Unsurprisingly, then, nothing material transpired immediately as the upshot of the March, and the marchers returned home disillusioned. Yet over time, views of its impact have changed. It can now be seen as part of the prompt to the growing awareness of regional inequality in Britain and has become a talisman. Unlike the shunning of the March by the Labour Party in 1936, the 1945 Labour Government adopted it as a metaphor for the harshness of unfettered capitalism and used Jarrow banners on its posters, and subsequent Labour leaders have associated themselves with the March.

JARROW MARCH

From 5th-31st October, 1936, 200 men from Jarrow, backed by their MP, Ellen Wilkinson, marched to London to protest over lack of employment opportunities.

Radical Tea Towel Co., *Route of Jarrow March, 1936* (1936) [RADICAL TEA TOWEL CO.] Ellen Wilkinson MP leading the March.

LOW FELL STATION

EASTERN AVENUE

TRIAL PIT NO. 4

M.H. EXISTING SOIL SEWER

FUTURE CHANNEL OF RIVER

EXISTING LAKE

ROAD NO. 8

TWENTY POUND CLOSE

PROBABLE FUTURE CHANNEL FOR RIVER

TRIAL PIT NO.

SPORTS GROUND

MAIN LAKE NORTH

ROAD

BRANCH DRAINS (SOIL & S.W. 9" DIA)

OPERATION OF WAGONWAY TO BE
PROVIDED FOR DURING CONSTRUCTION
OF ROAD. THIS PORTION OF ROAD TO BE
DEFERRED UNTIL WAGONWAY IS DIVERTED
(UNDER ANOTHER CONTRACT)

BRANCH DRAINS 9" DIA. S.W, 6" DIA SOIL

...AM COLLIERY WAGONWAY

PROBABLE FUTURE
DIVERSION OF BROOK

CULVERT (NOT IN THIS
CONTRACT)

ROAD NO. 1

ROAD

ROAD
TO NOTTINGHAM

COACH ROAD

SCALE 1/2500

FT 100 50 0   100   200   300   400   500   600   700   800   900   1000   1100   1200   1300   1400   1500   1600   1700   1800   1900   2000   2100   2200   2300   2400   2500 FT.

1936b

Team Valley: *an innovative industrial estate*

This is the best bit of news that has come to Gateshead for years . . . Too long has Tyneside had to rely on heavy industries . . . This scheme will go a long way to alleviate that position in the future.

Alderman White, quoted in F. Manders,
A History of Gateshead, 1973

The despondency that was felt by the Jarrow marchers was slightly misplaced because their efforts contributed to the growing sense that the state should tackle the economic and social effects of the country's increasing regional imbalance. Predating the March, an early reaction was the declaration of 'special areas' in an Act of 1934. This was the first in a long series of measures to tackle the problems faced by Britain's severely depressed areas, with the main places targeted by the Act being South Wales, west Cumberland, the central industrial belt of Scotland and parts of Tyneside. Two commissioners were appointed and given powers to develop ways of stimulating the economy of depressed areas, with an initial budget of £2 million to spend via local authorities, a sum that was increased by a further £2 million in 1936 and another £3.5 million in 1937.

One of the ways they chose to tackle economic issues was the creation of trading estates, modelled on the highly successful private-sector estate in Slough. In South Wales, Treforest Trading Estate was established close to Pontypridd, while in Scotland the Hillington Industrial Estate was opened in 1938 close to Glasgow (producing the first Merlin aircraft engines just before the start of the Second World War). On Tyneside, a North East Trading Estate company was established, and initially four potential sites were considered, one of which was Jarrow Slake. However, its chairman, Colonel

---

OPPOSITE: Sir Alexander Gibb & Partners, detail from *Team Valley Estate: General Plan* (1936) [GLH]
The labels 'future channel of river' denote the requirement to straighten and culvert large sections of the Team.

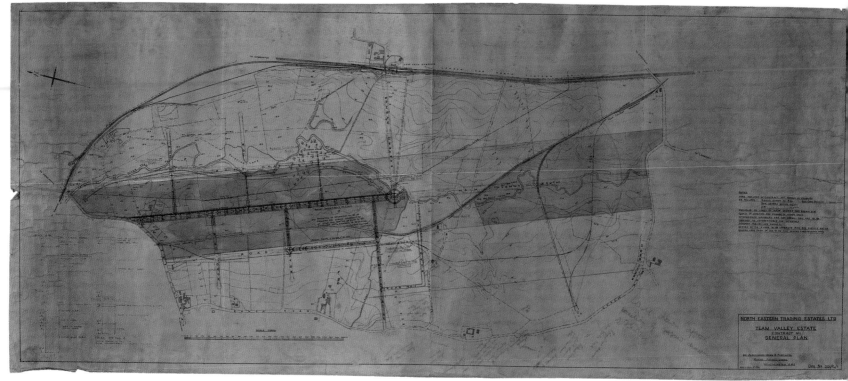

Alexander Gibb & Partners, *Team Valley Estate General Plan* (1936) [GLH]. Low Fell station was to be connected to the estate by Eastern Avenue, and was expected to play an important transport role, but was closed in 1952.

Appleyard, successfully argued for the valley of the River Team, a small stream that flows into the Tyne just upstream from Gateshead. Work on the site began in May 1936, with the first factory opening before the end of the year and with the estate being officially opened by the King and Queen in February 1939. In some ways Team Valley was a curious choice of location as it lay well to the west of the most deprived areas such as Jarrow and Hebburn, but it offered a conveniently flat site for modern factory construction.

The original site plan of the 700-acre estate was drawn up by Sir Alexander Gibb & Partners in September 1936, although the £800,000 contract for roads and services was awarded to George Wimpey & Co. the following month. Architectural design was carried out by W.G. (later Lord) Holford. The map shown here falls into the category of a utilitarian site plan where accurate and precise surveying is essential. The details are concerned with site engineering, road construction and the specifics of the drainage including the exact placement of manholes, as well as with the interim arrangements for existing infrastructure features whilst the site was under construction.

The pencilled annotations on this copy demonstrate that it was a 'working' document and provide some insight into how the plan was used on a day-to-day basis. The development was scheduled to take place in three stages with the northern (blue) section developed first, the middle (yellow) next and the southern area (green) last. The River Team itself, running through the centre of the estate and formerly described as 'a malodorous stream', was largely culverted and its boggy flood-plain stabilised with millions of tons of coal waste. The central spine of the estate was a 174-foot-wide dual carriageway, then the widest road in Britain, later called Kingsway and stretching two miles north–south.

The estate succeeded in its aim of attracting new firms: by May 1938 some 45 factories had been built, all but two of which were occupied; by June 1938 over 2,000 jobs had been created; and before the onset of war, the estate had 110 factories employing almost 4,000 people. However, these numbers were small substitute for the massive job losses created by the decline of the heavy industrial staples of the region. Indeed, most commentators – both at the time and subsequently – have argued that the estate failed to make significant inroads into the problems caused by high unemployment on Tyneside. Rather, they saw public works such as major house-building and road-improvement programmes as the saviours of Tyneside's workforce and, even more so, the government's pre-war rearmament drive. A single trading estate was never likely to turn around the fortunes of the area, but proponents of the estate could argue that the rearmament programme brought only short-term benefits and that the heavy-engineering jobs it created were not what Tyneside needed in the longer term.

The powers given initially to the two national commissioners enabled them to develop not only some economic leverage (such as through compulsory land purchase powers) but also a range of social initiatives, though they were originally precluded from assisting private companies. This restriction was removed in an Amendment of 1937, which also gave them new powers to lease factories and remit tax liabilities, so that the provisions of the legislation became much more consciously an attempt to tackle economic distress. In this, the Special Areas Commissioner for England and Wales, Sir Malcolm Stewart – an industrialist and philanthropist and chairman of the boards of both the London Brick Company and Portland Cement – played a significant part. He argued, unsurprisingly, that more money should be targeted to these areas but also, more radically, that there should be an embargo on new factory building in London. His ideas prompted the government in 1937 to appoint a Royal Commission on the Distribution of the Industrial Population under the chairmanship of Sir Anderson Montague-Barlow. It was the Barlow Report of 1940 that succeeded in developing the first real analysis of the 'regional problem' in Britain, in effect laying the foundation for the whole slew of acts that created Britain's urban and regional planning system. As the 1952 brochure shows, in the North East, 'Industrial Estates' (or, more correctly, 'Trading Estates') were a fundamental part of the regional planning strategy adopted post-war and designed to attract 'new' industries to modern premises strategically located near transport infrastructure.

Barlow's fundamental argument had been that regional deprivation arose from 'the structural effect'; in other words,

that the growth of more prosperous areas resulted overwhelmingly from their more favourable economic structure such as – at this time – light industries typified by electronics and consumer-based industries. The switch from industrial production to consumer goods had started in the late nineteenth century but, unlike Germany and the United States, Britain largely failed to make economic changes, instead relying heavily on its export trade and heavy industry base. Few parts of the country were more reliant on this increasingly outmoded economic model than the North East, and rearmament at the end of the 1930s merely exacerbated this. Team Valley was largely involved in light manufacturing and the argument that it had little impact on unemployment in Jarrow, a mere ten miles away, can be explained by some of the inherited characteristics of the area's staple industries. Female employment accounted for a substantial proportion of the new jobs at Team Valley, for example. Moreover, commuting even a short distance was a largely foreign concept for Tyneside shipworkers and redundant miners in County Durham's small pit villages.

In its own terms, the Team Valley industrial estate has been an undoubted success. In 2017, there were some 700 companies there, with some 20,000 people commuting in from across the region. But, in terms of creating a more radical shift in the nature of the North East's economy and its labour force, its impact was limited.

NE Trading Estates Ltd, *Industrial Estates in the North East* (c.1952) [AUTH] The widespread distribution of estates shows their significance in regional policy and the modernist architecture symbolises the break with previous industrial infrastructure.

Angaben nach den bis zum
vorhandenen Unterlagen

STADTPLAN SOUTH SHIELDS (Zusammendruck)

Sonderau
Ausgabe Nr. 2
Nur für den Dier

1942

Tyneside at war

Considering the importance of Tyneside to the war effort, it is surprising that the area was not bombed more intensively, but it did suffer substantial raids especially in late 1940 and throughout 1941. It was the sixteenth worst affected area in Britain.

A.W. Purdue, *Newcastle: The Biography*, 2011

The map of bombing targets around the mouth of the Tyne clearly demonstrates the thoroughness of German preparations for an aerial attack on Britain early in the Second World War. Specific military targets (*objektbildern*) were identified and numbered with great precision. Each map would be accompanied by a notebook with the corresponding details. This German research for cities and towns across Britain represents a staggering volume of preparation. The targets were presumably deemed to be strategic and, in the case of Tyneside, clearly include the many main docks, shipyards and industrial plants. Surprisingly, railway lines and junctions were not included, although North Shields railway station was. It is possible, however, that some details may have been mistranslated since the North Shields Workhouse, for example, is identified as a strategic target!

However, these targeting maps and their accompanying data are not a good guide to where bomb damage occurred because the number and type of bombs dropped show little relation to these targets, reinforcing the point that the Second World War impacted on civilians as well as the military. The detailed 'bombing' map of North Shields shows how a large number of bombs dropped between June 1940 and March 1943, presumably intended for Albert Edward Dock and Northumberland Dock, actually fell on the western residential part of North Shields itself. Many were also recorded as falling

OPPOSITE: *Luftwaffe Bombing Target map: River Tyne* (1942) [TWA] Target maps are believed to have been constructed from aerial photography which was then interpreted with the help of commercial directories.

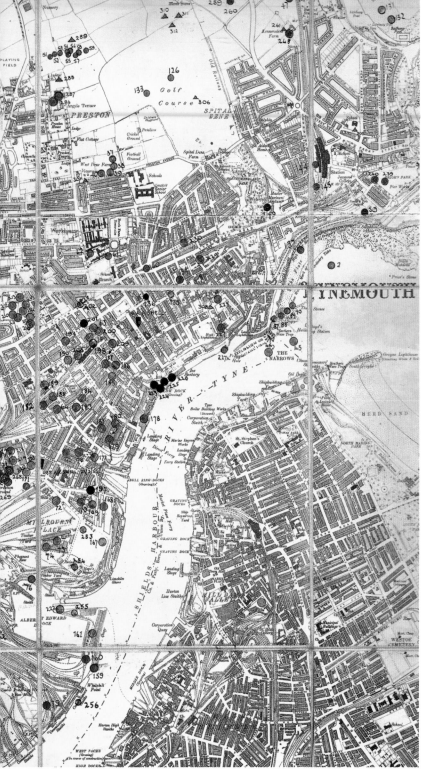

Tynemouth County Borough, detail from *Bombs dropped on Tynemouth and North Shields (June 1940–March 1943)* [NTCL] Red circles = High explosive bombs; Black circles = Unexploded bombs; Red triangles = Incendiary bombs.

into the sea off the coast between Tynemouth and Whitley Bay. In all, a total of 329 recorded bombing incidents in North Shields destroyed 166 properties and severely damaged another 1,307. The most tragic of these was the destruction of Wilkinson's lemonade factory (hardly a strategic military target) in May 1941, causing the building and its machinery to collapse into the air-raid shelter in its basement, killing 106 people.

This disaster had national reverberations since, within days, the maximum number of people allowed to occupy basement shelters was reduced to 50. The shelter beneath the lemonade factory was badly sited and certainly grossly overcrowded, but the location of air-raid shelters was always a trade-off between rapid ease of access and efficiency of protection. With the realisation that air raids could last a long time, bigger and more heavily protected sites were sought. In Newcastle the two best examples were the conversion of the Victoria Tunnel (which was originally built to carry coal from Spital Tongues to staithes on the Tyne at the mouth of the Ouseburn) and the opening of the culvert over Ouseburn just upstream from the Tyne. Seven new entrances were created along the length of the Victoria Tunnel (numbered on the map) which had a capacity of 9,000. The Ouseburn culvert, built of ferro-concrete, could take 3,000 people and had its own sick bay. These sites are now popular heritage experiences, with underground tours that are invariably oversubscribed.

Remnants of military infrastructure, too, can still be seen on Tyneside, for example the former coastal batteries at Tynemouth Castle, Clifford's Fort in North Shields and Frenchman's Point in South Shields. Other defensive measures such as barrage balloons, tank traps on the beaches and pill boxes, although an integral part of the wartime landscape, have now almost entirely disappeared.

Although war was not declared until September 1939, Tyneside was not entirely in a state of unpreparedness. By March 1939, 1,800 Anderson shelters had been built in the Newcastle area and 9,000 more were under construction. However, the demands of total war took some time and considerable dispute before being accepted, particularly in the case of the requisition of buildings. Responding to the possibility that premises could be commandeered for the purposes of air-raid and first-aid shelters, the Methodist Church, for example, circulated its property managers with the instruction: 'No building devoted to public worship is to be claimed by the national or local authorities for any emergency purpose.' Conflict was apparent in other spheres such as the Ministry of Transport's attempt to claim Newcastle Quayside's five-ton crane, which was strenuously resisted by the City Engineer: 'All the cranes on the Quay are fully engaged on . . . rapid unloading to facilitate the quick turn round of ships. I regret that I cannot approve of the 5-ton crane being transferred to the War Office.'

More generally, the war had significant impact on the region's economy. The Elswick factory (by 1939 owned by Vickers-Armstrong) specialised in tank manufacture throughout the war. Shipbuilding received a welcome boost and the output of cargo ships and naval vessels was impressive. For example, Swan Hunters of Wallsend built 81 warships, including the battleship HMS *Anson*, two aircraft carriers, three cruisers and 28 destroyers. The Walker Naval Yard of Vickers-Armstrong built HMS *King George V*, four carriers, three cruisers, 24 destroyers and 16 submarines. R. & W. Hawthorn Leslie built 41 naval vessels and were particularly notable for their repairs of damaged vessels, most famously HMS *Kelly* in 1940. Less well known is the vital role that Tyneside firms played in the manufacture of the Mulberry Harbour components which were so fundamental to the success of the D-Day landings in France and the subsequent supply logistics. However, it has been suggested that the long-term impact of the wartime upturn in shipbuilding activity was to perform a disservice to the Tyneside economy, as the war simply offered a temporary reprieve for traditional heavy industries in long-term decline. It also further reinforced the reliance of large firms on government contracts. Ironically, what could have been the basis for a more modern industry – aircraft manufacture – was directed away from Tyneside on the grounds that the area was vulnerable to air attack.

Despite the impressive statistics on output, any notion that

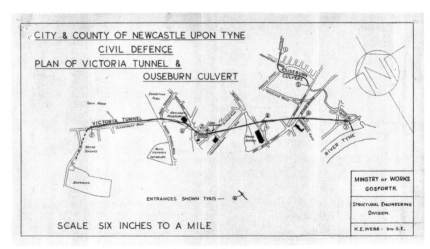

K.E. Webb, *Plan of Victoria Tunnel and Ouseburn Culvert* (1939) [SANT] Entrance number 6 can still be seen in Claremont Road, outside the Great North Museum.

Tyneside workers were consistently united in their war effort at the cost of all other considerations is simply a myth. Strikes were a frequent occurrence and actually increased during the war years on Tyneside, as nationally. The vital shipbuilding, engineering and steel production industries had been responsible for 7 per cent of days lost in manufacturing industry strikes in the 1927–38 period, a proportion that increased to 32.2 per cent in the first three years of the war. Any sense that 'we are all in this together' was also severely impacted by the exposure of corruption and profiteering within the Air Raid Precautions (ARP) system at the highest level, notably involving Newcastle's Chief Constable Crawley who headed the ARP structure and was forced to resign in disgrace, along with Councillor Embleton.

The impact on the region's social structure was rather more subtle. As in the First World War, one obvious change was the employment of women in heavy industry. For example, at Palmers old yard at Hebburn, 350 women were employed in ship repair and conversion. Quantitatively far more important were the jobs offered to women by clothing firms relocating from the South East to the North East in search of cheap female labour. The employment of women in heavy industry was a short-lived phenomenon rather than a long-term trend. Ultimately, Tyneside did share in the national trends of change in gender relations, but there is little evidence in the immediate post-war years of any erosion of the region's patriarchal reputation.

1947

The sentinel towns in the twentieth century

Tynemouth, with the ruined castle and priory standing on a cliff overlooking the North Sea and the entrance to the Tyne, is like a vision of another world.

G. Moorhouse,
Companion into Northumberland, 1953

The map of the mouth of the Tyne at the scale of one inch to one mile (1:63,360) is a fine example of the aptly named New Popular edition published in the mid twentieth century. This series of maps is probably the most familiar of maps to many readers, certainly those who grew up in an era when paper maps were the principal mode of place and wayfinding, rather than a digital device. They are likely to have been that generation's first encounter with a properly surveyed map and probably were a standard feature of school geography lessons.

This series was the first to use the metric National Grid introduced to facilitate the location of places and phenomena by co-ordinate referencing. Colour had been introduced into the 'Old Series' 1:63,360 maps in 1897 but featured much more strongly in the 'Popular' edition. Relief was represented by contours instead of hachuring and there was a deliberately enhanced attempt to depict non-physical features on the maps. These maps also demonstrate a further movement away from the military origins of the Ordnance Survey where the representation of terrain and physical features were paramount. Built-up areas, a sophisticated road classification (a product of the huge increase in motorised road transport), the detailed naming of places and the identification of most infrastructural features were the dominant characteristics of the Popular edition, all presented in striking colour.

OPPOSITE: Ordnance Survey, *Coastal Detail of Sheet 78, Newcastle upon Tyne* (1947) [NLS]
The New Popular edition provides a record of Britain immediately after the Second World War
and on the brink of massive social, economic and environmental change.

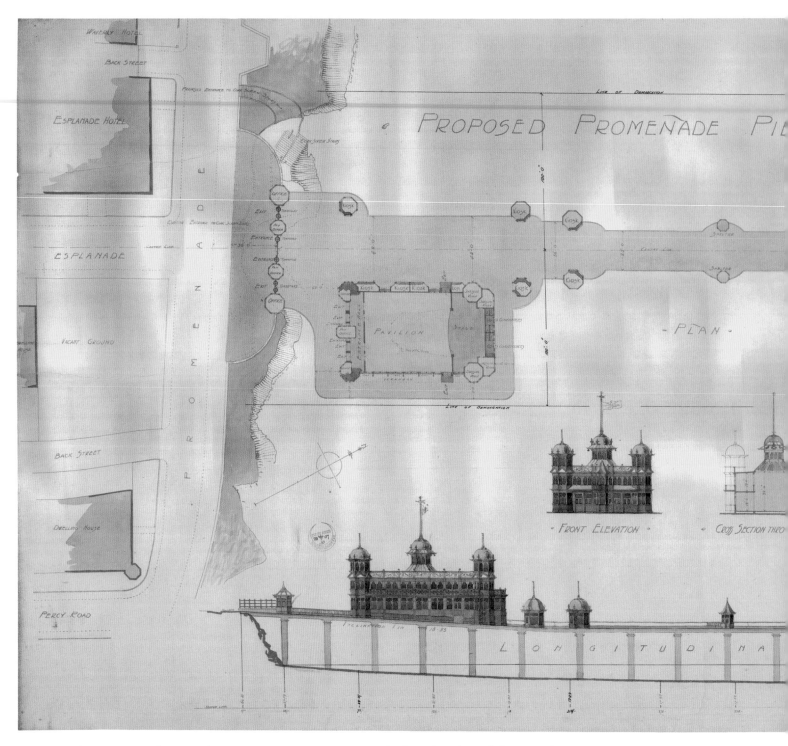

White & Stephenson, *Proposed promenade pier at Whitley Bay* (1907) [N.EST] The pier was planned opposite the Esplanade leading directly from Whitley Bay station to the new upper promenade with the Corkscrew Stairs, a landscaped feature created in 1893.

The broad outline of the immediate post-war settlements on the north and south banks at the mouth of the Tyne are shown as still predominantly clinging to the riverside. But North and South Shields developed rather differently in the

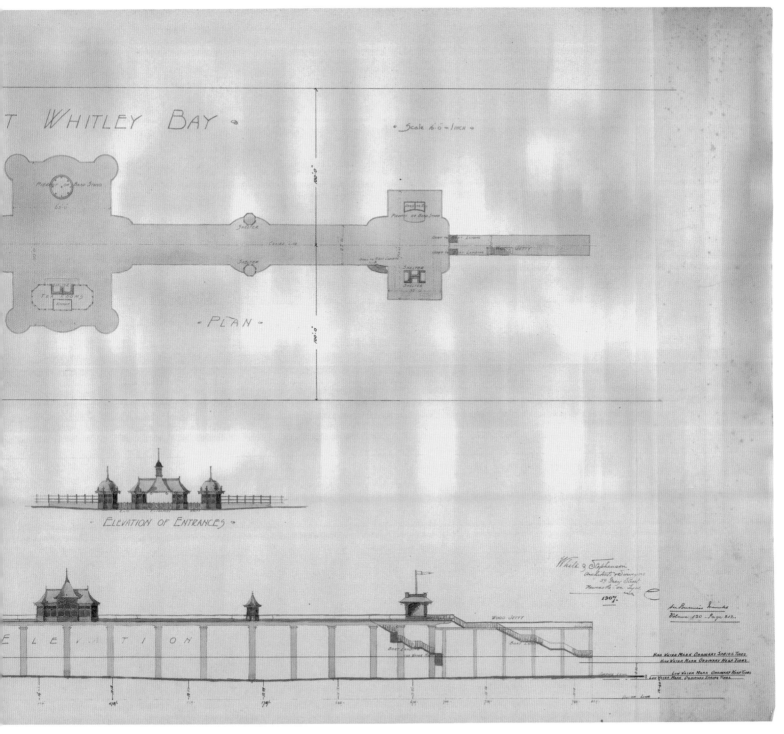

twentieth century. Both banks of the river were interspersed with terraced houses, many of them 'Tyneside flats' and home to shipworkers, metalworkers and other skilled and semi-skilled households. However, on the north bank of the river, two distinct settlement areas had emerged: an industrial town comprising that part of North Shields lying along the banks of the river and upstream towards Wallsend; and a much wealthier suburban area comprising Tynemouth and the

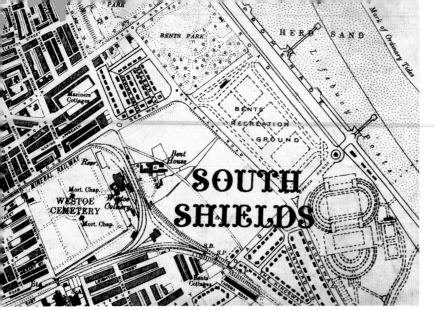

D.M. O'Herlihy, detail of the south-east corner of *County Borough of Tynemouth Map* (1948) [AUTH] This 1948 map shows the concentric circular arrangement of the wartime barrage balloon site, bottom right.

plateau area onto which North Shields had spread. South Shields remained a uniformly working town driven by its industry. The narrow stretches on both sides of the Tyne continued to have industry along the river, especially shipbuilding and ship repairing, with Smiths Docks dominating North Shields and Readheads dominating South Shields. The fish quay at North Shields was home to a substantial fleet of trawlers which generated a healthy trade in fish, and even as late as the 1950s huge whale boats landed their cargo at the quay – occasions of great excitement for onlookers and a spectacle that today would doubtless be greeted by protestors.

Significant change had been taking place through the twentieth century, even in formerly 'professional' areas such as Dockwray Square on the plateau area of North Shields whose status was gradually eroded. In fact, the Square had been poorly provided with water and suffered from drainage difficulties, problems that were never adequately resolved, with many of its houses becoming tenements. They were eventually demolished and replaced in the 1960s with flats built around a communal recreational area, with these in turn being replaced in the 1980s by private housing including a new square. The latter has a statue of Stan Laurel standing at its centre to commemorate his living in a house in Dockwray Square earlier in the century. Meanwhile, Tynemouth increasingly developed as a fashionable area attracting professional families, many of

whom worked in Newcastle.

South Shields remained an overwhelmingly industrial town with just a few residential areas for professional families. Most of the latter chose to live well outside the town in small villages to the south such as Westoe, Cleadon, and even Whitburn which lies halfway to Sunderland. The difference continues to be apparent more recently in the pattern of commuting to work. In 2011 over 12 per cent of North Tyneside's population commuted to Newcastle, while for South Tyneside it was less than 5 per cent, with slightly more commuting to Sunderland.

A major development in the twentieth century was attempts by both towns to attract tourism, although Tynemouth's tourist trade was significantly greater than that of South Shields. Tynemouth's draws were obvious, not least the castle and priory, and its beaches had long attracted tourists, with sea bathing being fashionable here from the late eighteenth century. The main beach, Long Sands, stretches north to the attractive village of Cullercoats, while further north again Whitley Bay is also a desirable residential area. Although Whitley Bay's attempt to develop a promenade pier with associated pavilion was not successful, the Spanish City, with its distinctive dome and twin towers, was one of the attractions on offer. It was initially established in 1910 as a concert hall and restaurant, and later a ballroom and funfair were added. It has recently been refurbished as a restaurant complex.

South Shields attracts tourists on a much smaller scale. This is despite the seaside attractions that it offers. South of the river mouth, Ocean Beach stretches southwards with fine sand, an amusement park and two adjacent large parks (South Marine with a boating lake and Bents Exhibition Park), created in a period of depression using formerly unemployed labour. Changing functions and priorities are reflected in a remarkable early example of industrial land transformation affecting 140 acres as these parks were developed on what had been derelict former industrial land (glass works) and ballast hills, having been bought relatively cheaply from the Ecclesiastical Commissioners. The Borough Surveyor, Matthew Hall, designed North Marine Park and won national acclaim from the *Gardener's Calendar* of 1886; he had 'transformed the barren waste into a landscape which will form a most healthful ornament to the town'.

Further south, cliffs of magnesian limestone are studded with dramatic caves. Marsden Bay, shown at the bottom right-hand corner of the lead map, is a second major stretch of sandy beach leading to the iconic Marsden Rock and the Grotto, a cave bar (and later a restaurant) partly carved from inside the cliffs. It had been blasted at the end of the nineteenth century by a Durham lead miner who enlarged it from a small cave into a larger 'house', and it was progressively expanded into a pub at the base of the cliffs, accessed either down rickety wooden steps or by a penny lift.

Yet even in Marsden, the presence of industry was long apparent. Close to the cliffs, the South Shields, Marsden and Whitburn Colliery Railway ran from Whitburn colliery to South Shields. Construction of the line began in 1878, opening for mineral traffic in 1879, and by 1888 passenger services were given permission to run along the line, with trains consisting of three coaches, plus a guard's van. Known as 'the Marsden rattler', reflecting its somewhat ancient rolling stock, the whistle of the engine and its belching smoke were a familiar sound and sight above the cliffs of Marsden. The line stopped carrying passengers in 1953 but continued to carry coal until the colliery's closure in 1968.

Since the turn of the millennium, the riverside sections of both Shields have undergone a remarkable transformation. Large-scale clearance of former industrial areas has created sites for modern residential development with a 'river view' carrying a premium. Economic change is encapsulated by the Royal Quays development in North Shields, shown on the accompanying map. This 200-acre site contains 1,200 new homes (including around 25 per cent for rent) and a retail outlet, and the former Albert Edward Dock is now a marina. Across the river, the South Shields 365: Riverside project in the Holborn area is similarly recreating a new residential and commercial neighbourhood on formerly derelict land. After a lengthy and complex process of site assembly, nearly 400 new dwellings are to be built, along with an office development creating 500 jobs and a small business complex. The Grade II

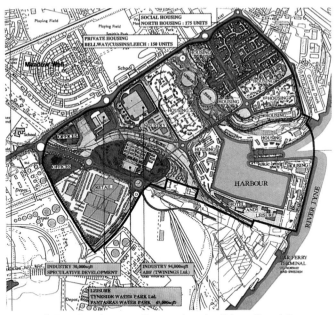

Tyne and Wear Development Corporation, detail from *Royal Quays Infrastructure Plan to 1994: North Tyneside* (c.1998) [NERA] The 'harbour' is part of the former Albert Edward Dock and is now a marina.

listed Customs House building (dating from 1863–64) is to be used as the focus for an emerging cultural quarter.

The sentinel role that the twin settlements have played since Roman times is now more apparent. At South Shields, *Arbeia* was a large Roman fort which protected the main sea route to Hadrian's Wall and was the maritime supply fort for the Wall. Although first excavated in the 1870s, it had long been an insignificant feature. However, nineteenth century buildings on the site were cleared in the 1970s and the fort has partly been reconstructed with a large Roman gatehouse, barracks and officers' dwelling. The gatehouse acts in part as a museum. Complementing *Arbeia*, on the north of the Tyne, the Roman fort of *Segedunum* at Wallsend formed the eastern terminus of the Wall and has been exhaustively excavated. The site now displays the remains of the fort's original foundation, as well as a reconstructed bathhouse, a museum and a large observation tower. These two forts now effectively round off the circle of history by taking us back to the beginnings of the significant roles played by the river and Hadrian's Wall.

1949

Jesmond and Gosforth: suburban development

The professional and business classes who could afford to make their homes farther afield moved into Jesmond, Fenham . . . and 'Bulman's Village' as Gosforth was called before 1914.

S. Middlebrook, *Newcastle upon Tyne.*
Its Growth and Achievement, 1950

The position and layout of Newcastle's predominantly middle-class suburbs of Jesmond and Gosforth is shown in the 1949 map. The development of both areas had similarities. Unlike many suburbs, neither developed from a substantial medieval village core, nor were they the result of conscious planning to accommodate the ideal middle-class lifestyle and community. Their growth was what F.M.L. Thompson called 'an intricate mosaic of building estates and developments', in fact a palimpsest of phases of development and accompanying social and physical characteristics. Nevertheless, although taking place at different times and producing different components in the built environment, the development of both Jesmond and Gosforth followed a similar sequence.

Jesmond experienced colonisation by the 'new' commercial and industrial entrepreneurial class (most notably Sir William Armstrong) in the early part of the nineteenth century, and Gosforth somewhat later. In both areas an initial phase of dispersed colonisation by large, detached houses in substantial grounds, almost 'mini estates' (shown well in the detail from Oliver's 1844 map of Newcastle), was later followed by land sales to create further scattered substantial 'villa' properties. A later process of subdivision created new street lines with associated building plots for groups of smaller, yet still substantial, residences. In Jesmond in particular, many of these were large, terraced houses, although in both Jesmond and Gosforth the

OPPOSITE: Andrew Reid, *Map of Gosforth & Jesmond Suburbs* (1949) [AUTH] Andrew Reid & Co., successor to M. & M.W. Lambert, was a hugely important provider of maps of Newcastle and its region throughout the later nineteenth and twentieth centuries.

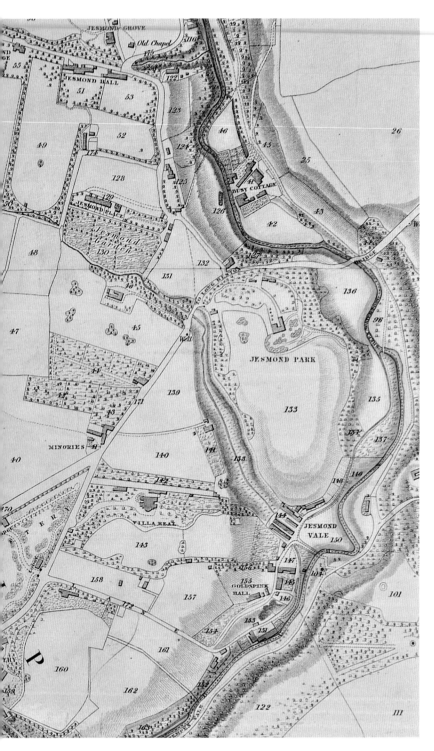

scattered 'country villas', which had imitated on a small scale the residences of the landed gentry, remained an integral part of the landscape character. A fourth phase saw denser development of smaller, mainly terraced properties, demand being fuelled by the growth of lower grade 'white collar' employment and the emergence of building societies. In contrast to previous phases, although a domestic servant may have been employed, there was no room to provide accommodation. However, the subsequent development of the two areas has diverged considerably, partly due to their different distances from the city centre.

In Jesmond the ambience was largely set in 1824 by John Dobson's magnificent entrance to Jesmond cemetery, located on the bottom left of Oliver's 1844 map. The area of initial residential colonisation was between the cemetery and the Dene, an area which offered considerable amenity value with its spacious wooded environment and its location at a comfortable distance from the high density and increasingly polluted environment of central Newcastle.

A subsequent phase of suburban development in Jesmond, from the 1870s onwards, was located to the west of this initial colonisation. The developments varied in style and grandeur, but each took place under strict covenants relating to building materials, size, facilities and use. The latter invariably excluded premises licenced for the sale of alcohol. One of the grandest developments was the section east of Osborne Road between Jesmond Road and Clayton Road, which was undertaken by James Pears Archbold between 1863 and 1875. With its large villas, the area became the residence of people like the department store owner T.H. Bainbridge, the Wallsend shipbuilder G.B. Hunter and the shipowners Walter Runciman and Sir Arthur Munro Sutherland. In contrast, later developments further to the west, adjacent to the railway line, were of more modest proportions and the character of the area began to change substantially.

Thomas Oliver, detail of Jesmond from a *Plan of Newcastle* (1844) [NCA] The grand villas of Jesmond Hall, Jesmond Grove, Jesmond Park, Villa Real and Goldspink Hall set the early exclusive tone for Jesmond.

Thomas Hills Forsyth, chairman of the Rock Building Society from 1880 until his death in 1905, built up West Jesmond from 1894 and included streets of terraced Tyneside flats. The plan of 1895 shows the layout of streets on Forsyth's land to the east of the railway line and his development was clearly related to the proposed opening, in 1904, of West Jesmond station. The piecemeal nature of building, albeit within an overall structure laid down by Forsyth, is indicated with the areas shaded red already under development, including St Hilda's church in Thornleigh Road. Plots were sold to small-scale builders who constructed small groups of houses or flats either for a rental income or sale to small scale investors. Forsyth continued his development on the western side of the railway line and this part of Jesmond was framed on its western edge by three large Victorian charitable institutions facing the Town Moor – an Institute for the Deaf and Dumb, a Hospital for Sick Children and an orphanage. The presence of these institutions helped to create an appropriate atmosphere for the later development of several private schools in this part of Jesmond.

The core of Gosforth (known initially as Bulman's Village) developed on what is now the High Street, with a number of large detached properties being built. Dominant was Coxlodge Hall, built in 1796 to the west of the main road and inherited by Job James Bulman in 1818. Bulman, who had made his fortune in India and was a partner in Lambton's bank of Newcastle, owned most of the land around the village core and sold off plots for the construction of substantial villas. The western side of the Great North Road was relatively undeveloped in the mid nineteenth century but, as shown on the map, by 1898 a number of large detached stone houses had been built, especially adjacent to the amenity value of the Town Moor and with access to Kenton Road. In 1873 the local historian Richard Welford described the outcome: 'Large numbers of snug suburban residences are springing up, built in almost every variety of residential architecture, from Gothic villa to modest cottage *ornee*.' In the early twentieth century development accelerated and existing east–west streets were intersected by a series of north–south streets which created smaller

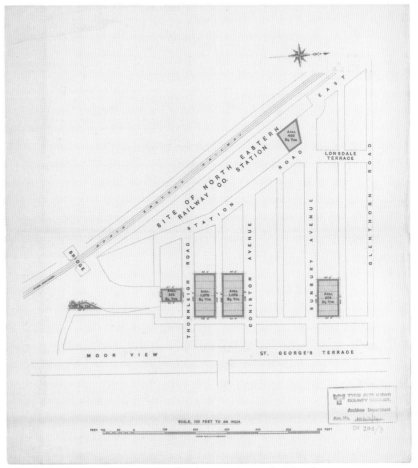

Estate development: T.H. Forsyth's plan for West Jesmond with railway station (1895) [TWA] T.H. Forsyth also developed an extensive area to the west, between the railway line and the Great North Road.

(although still extensive) plots. The result overall was a pattern of substantial houses, although with variations in house type dictated by plot size. At this time, decorated terraces and large semi-detached houses in the Queen Anne Revival style predominated, exemplified by Graham Park Road which was developed on part of the Coxlodge Estate. The main architect was William Hope (1862–1907) whose work was strongly influenced by Norman Shaw, the architect of Cragside, Sir William Armstrong's house near Rothbury.

The development of the Ashburton Estate in the early

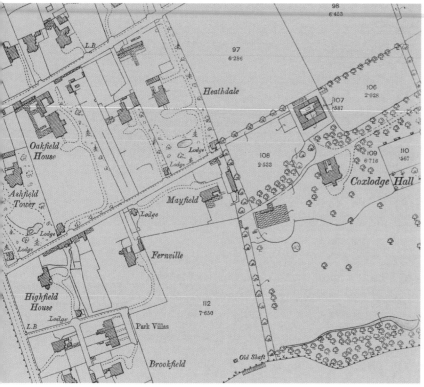

Ordnance Survey, *Large detached villas at Gosforth* (1898) [AUTH]
Coxlodge Hall (1796) was demolished in 1939 and its grounds
developed for large detached houses. The stable block to the north of
the Hall remains and is now luxury flats. The driveway to the right
survives in the modern street name The Drive.

twentieth century produced a different grain. The sale of the
estate in 1883 initially led to institutional development,
including the Roman Catholic cemetery, but streets of substantial terraced houses were planned from around 1900. Being
somewhat 'historic', the inset plan of Gosforth gives a slightly
misleading impression of an exclusive, low-density environment as higher density terraced development was already
underway as evidenced by the Ashburton Estate itself. As in
Jesmond, individual plots or small groups of plots were sold
to small-scale builders who would then either sell on the
completed properties to property investors or rent out the
dwellings themselves.

Although often coupled together as archetypical middle-

class suburbs, Jesmond and Gosforth exhibit differences in
detail, and their more modern trajectory of development has
served to sharpen these. Both are substantially covered by
Conservation Area status, but the problems they face are
different. Jesmond's proximity to the city centre and the two
universities has led to a massive influx of students and young
professionals, leading to multiple occupancy of former large
family houses and consequent threats to their visual architectural integrity. The potential for conflict between older permanent residents and younger temporary inhabitants is now
considerable, exacerbated by the conversion of former hotels
in Osborne Road into bars and clubs. Jesmond's location has
also led to a colonisation of large properties by commercial
functions (Greggs, the highly successful bakery, has its
headquarters in one of the former large villas on Clayton Road
East).

Gosforth has not experienced the same colonisation by
students as Jesmond, but the busy High Street shopping area
creates its own spill-over problems including traffic management issues. So too does the office area of Regent Centre to
the north where small businesses, functionally related to the
offices, represent a threat to residential properties. In addition,
modern in-fill developments have changed the traditional
character of some streets, as have inappropriate extensions to
existing properties. A similar impact has resulted from the loss
of front gardens and their walls in order to create hard standing
for the parking of vehicles. The loss of green open space,
including trees, is also eroding one of the key features of the
area.

As a footnote on middle-class suburban life on Tyneside,
the Russian author and early Bolshevik, Yevgeny Zamyatin,
was living in Sanderson Road, Jesmond, in 1916 and passed
scathing comment on the conformity of Jesmond residents:
'Sporting identical canes and identical top hats, the Sunday
gentlemen strolled in dignified fashion along the streets and
greeted their doubles.' If Jesmond and Gosforth lacked variety
in 1916, the same cannot be said a century later. Both areas,
once the premier locations of the wealthy, have inevitably
changed their character as Newcastle has expanded. The

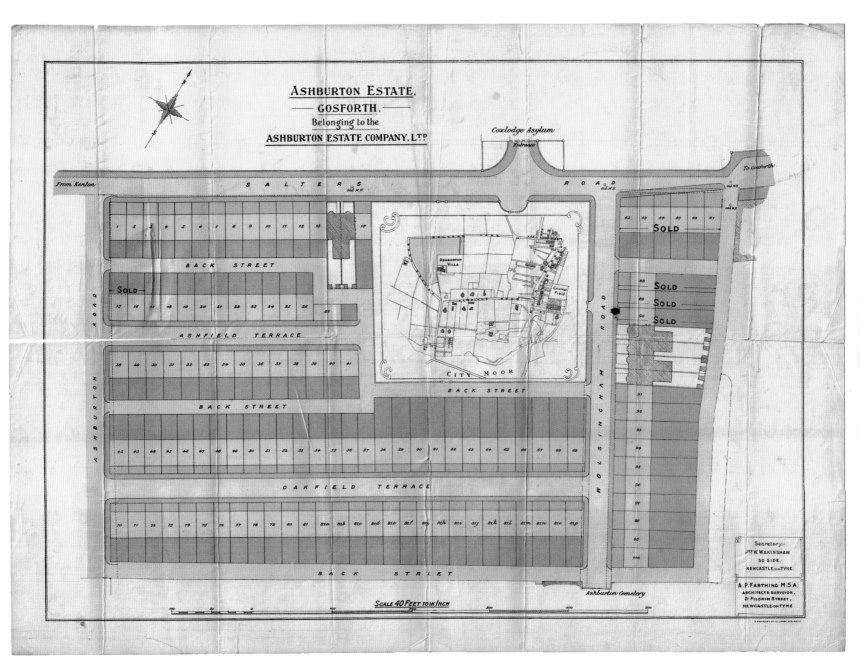

suburban ideal is now to be found outside the city boundaries, archetypically in Darras Hall next to Ponteland – home to several Newcastle United footballers – and up the Tyne Valley in Wylam, Corbridge and beyond.

*Development plan of Ashburton Estate, Gosforth (c.*1900*) [*TWA*] The residents of the Ashburton Estate complained in* 1916 *that the promised tramway along Salters Road had not yet arrived.*

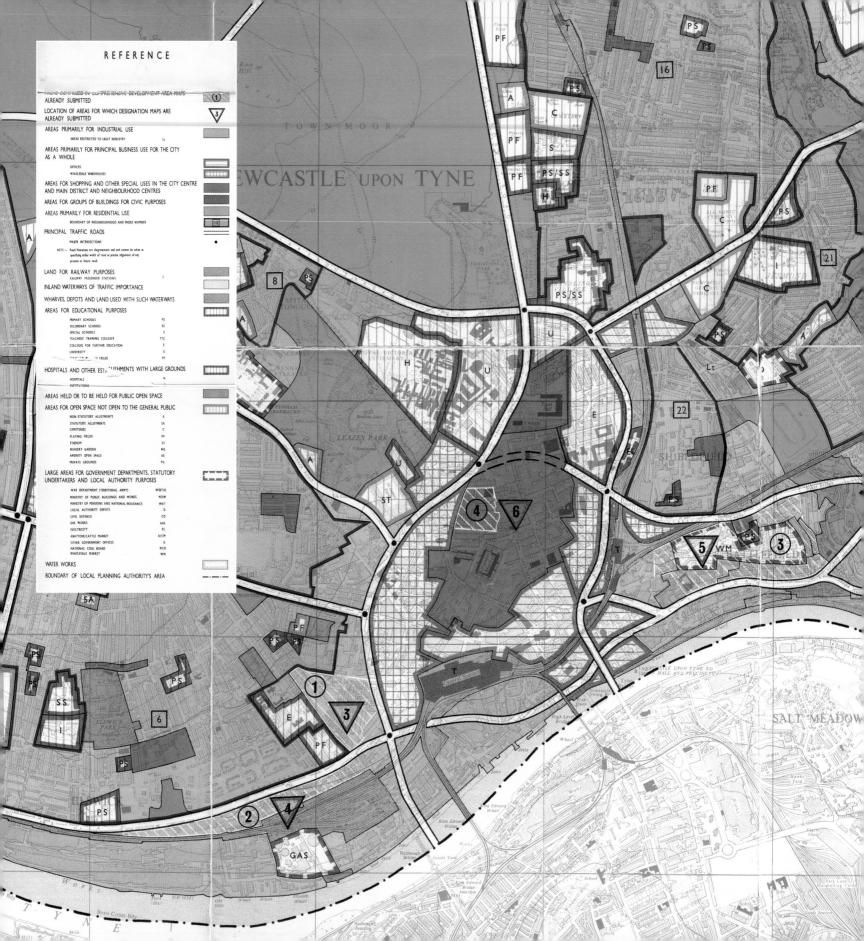

REFERENCE

AREAS COMPRISED IN COMPREHENSIVE DEVELOPMENT AREA MAPS
ALREADY SUBMITTED

LOCATION OF AREAS FOR WHICH DESIGNATION MAPS ARE
ALREADY SUBMITTED

AREAS PRIMARILY FOR INDUSTRIAL USE

    AREAS RESTRICTED TO LIGHT INDUSTRY   Lt

AREAS PRIMARILY FOR PRINCIPAL BUSINESS USE FOR THE CITY
AS A WHOLE

    OFFICES
    WHOLESALE WAREHOUSES

AREAS FOR SHOPPING AND OTHER SPECIAL USES IN THE CITY CENTRE
AND MAIN DISTRICT AND NEIGHBOURHOOD CENTRES

AREAS FOR GROUPS OF BUILDINGS FOR CIVIC PURPOSES

AREAS PRIMARILY FOR RESIDENTIAL USE

    BOUNDARY OF NEIGHBOURHOOD AND INDEX NUMBER   12

PRINCIPAL TRAFFIC ROADS

    MAJOR INTERSECTIONS

    NOTE:– Road Notations are diagrammatic and cannot be taken as
    specifying either width of road or precise alignment of any
    present or future road.

LAND FOR RAILWAY PURPOSES

    RAILWAY PASSENGER STATIONS   T

INLAND WATERWAYS OF TRAFFIC IMPORTANCE

WHARVES, DEPOTS AND LAND USED WITH SUCH WATERWAYS

AREAS FOR EDUCATIONAL PURPOSES

    PRIMARY SCHOOLS   PS
    SECONDARY SCHOOLS   SS
    SPECIAL SCHOOLS   S
    TEACHERS' TRAINING COLLEGES   TTC
    COLLEGES FOR FURTHER EDUCATION   E
    UNIVERSITY   U
    UNIVERSITY PLAYING FIELDS   PF

HOSPITALS AND OTHER ESTABLISHMENTS WITH LARGE GROUNDS

    HOSPITALS   H
    INSTITUTIONS

AREAS HELD OR TO BE HELD FOR PUBLIC OPEN SPACE

AREAS FOR OPEN SPACE NOT OPEN TO THE GENERAL PUBLIC

    NON-STATUTORY ALLOTMENTS   A
    STATUTORY ALLOTMENTS   SA
    CEMETERIES   C
    PLAYING FIELDS   PF
    STADIUM   ST
    NURSERY GARDEN   NG
    AMENITY OPEN SPACE   AS
    PRIVATE GROUNDS   PG

LARGE AREAS FOR GOVERNMENT DEPARTMENTS, STATUTORY
UNDERTAKERS AND LOCAL AUTHORITY PURPOSES

    WAR DEPARTMENT (TERRITORIAL ARMY)   WD(TA)
    MINISTRY OF PUBLIC BUILDINGS AND WORKS   HOW
    MINISTRY OF PENSIONS AND NATIONAL INSURANCE   MNT
    LOCAL AUTHORITY DEPOTS   D
    CIVIL DEFENCE   CD
    GAS WORKS   GAS
    ELECTRICITY   EL
    ABATTOIR/CATTLE MARKET   A/CM
    OTHER GOVERNMENT OFFICES   G
    NATIONAL COAL BOARD   NCB
    WHOLESALE MARKET   WM

WATER WORKS

BOUNDARY OF LOCAL PLANNING AUTHORITY'S AREA

1963

'Brasilia of the North': Smith, Burns and the halcyon days of post-war town planning

In Newcastle . . . planning is an all-out effort to abolish the past and manufacture the future.

J.G. Davies, *The Evangelistic Bureaucrat*, 1972

The term 'Brasilia of the North' was famously coined by T. Dan Smith, the leader of Newcastle City Council, to capture the image of the 'new' Newcastle that he attempted to develop in the 1960s. Its resonance at the time related to the highly publicised modernist architecture and planning associated with the development of Brasilia as a completely new city and capital of Brazil. The intention was to emphasise a complete break from the past. In his 1970 autobiography, Smith wrote: 'In Newcastle I wanted to see the creation of a twentieth century equivalent of Dobson's masterpiece, and its integration into the historic framework of the city.'

To bring his vision to fruition, Smith recruited a leading planner, Wilfred Burns, to direct what was one of Britain's first free-standing Planning Departments. Becoming a powerful advocate of modernist planning, Burns aimed to demolish almost one-quarter of Newcastle's older housing and replace its often mixed land uses with monolithic single uses. His 1963 plan also addressed the issue of how best to resolve the conflict between cars and pedestrians by proposing to enclose the city centre within an urban motorway box. In the shopping area and university precinct, traffic and pedestrians would be separated, with traffic remaining at ground level while pedestrians would circulate at first-floor level. The plan was initially warmly welcomed, but over time it increasingly ran into opposition, especially from conservationist lobbies. Even so, the plan led to many of the Georgian and Victorian buildings being demolished, not least with more than 50 listed buildings making way for the new indoor Eldon Shopping Centre, numbered 4 on the 1963

OPPOSITE: Newcastle City Council, detail of *Development Plan Review, 1963* (1963) [AUTH]
One account of Newcastle's post-war planning policy was headed 'A city fit for cars to park in?'

Newcastle City Council, *Photo of 3D Plan of Proposed Redevelopment of City Centre* (1966) [NCL] Wilfred Burns himself described the redevelopment of Newcastle city centre as 'an awesomely ambitious scheme'.

map (although its area later increased substantially).

Major losses from the Dobson/Grainger era were the Royal Exchange and most of old Eldon Square. The Royal Exchange had never been a great commercial success, given its off-centre location, but it was a fine building. Its demolition was due to the creation of the Central Motorway East (CME), the one section of the encircling road system that did materialise. The CME closely followed the line of the medieval wall, abutting the eighteenth-century Holy Jesus Hospital and leaving that building hunkered in a gloomy and noisy cavernous trough. Where the motorway left the Tyne Bridge, a large new round-about was constructed, and Swan House was built to commemorate Joseph Swan. It was an unimaginatively designed multi-storey office block that glowered down on the Bridge and has now been converted into residential apartments and 'night life' functions. The intention was to demolish and rebuild the Royal Exchange in a slightly different location. The ashlar stones were numbered so that they could be reconstructed, but the stones either deteriorated or – not surprisingly – the chalk numbering was washed off by rain. The reconstruction never took place and the whole building was finally demolished in 1969.

Eldon Square was one of Dobson's jewels. The architectural critic Nicholas Pevsner compared it favourably with Nash's terraces flanking Regent's Park in London. If the Exchange was a loss, the destruction of Eldon Square was a tragedy, lambasted by Christopher Booker in 1978 as 'perhaps the greatest single example of architectural vandalism in Britain since the war'.

Smith had hoped to see a tower designed by the Danish architect Arne Jacobsen to sit in the old square, but that ambition came to nought. Instead, two sides of the square were demolished to make way for the new shopping centre, with just its eastern wing surviving, this being on the grounds that its demolition would have destroyed a non-conformist chapel.

The powerful partnership of Smith and Burns has generally been demonised for the damage done to the Dobson/Grainger townscape. Views of Smith have doubtless been coloured by his imprisonment for his role in the 'Poulson affair' that involved bribing local councils to attract business, but the reality for both men is a little more complex. Burns may well have been an advocate of comprehensive planning driven by a master-planning approach, yet he included four 'preservation' areas (Grey Street, Grainger Street and the Grainger markets; the St John's Church area; the cathedral/castle area; and the Guildhall area), and this can be seen as pioneering, predating, as it did, the Civic Amenities Act of 1967 which introduced formal Conservation Areas. Smith's record can also show him in a rather more nuanced light. He spent much time discussing issues with architects and academics at the University, and he successfully fought to preserve the integrity of Grey Street, 'that superb testimony to Dobson, which I was determined to keep sacrosanct. The Midland Bank wanted to insert a piece of modern architecture. So I stuck my neck out and said No. I was accused of carrying a torch for ersatz Dobson.'

Even so, there is little doubt that the proposals that came

Ordnance Survey, *Terraced housing street layout in Byker, 1913* (1913) [AUTH] Unlike the west end of the city, Byker's terraces were oriented mainly in an east–west direction.

Street layout as designed by Ralph Erskine, Byker 2021 (2021) [OSM] An informal 'drop in' centre was established by Vernon Gracie, the project architect, in an empty former undertaker's shop.

from Burns and Smith, if fully implemented, would have emasculated the city. Growing resentment at the loss of the old was compounded by dissatisfaction with the quality of what replaced it. It was a happy accident that the lapse of time allowed mounting opposition which prevented the plan's full implementation, especially the completion of the motorway box. A recent history of the city (F.W. Purdue: *Newcastle: The Biography* (2011)) noted the longer-term repercussions of this era: 'From the early 1980s a reaction against the planning and architectural vision of the post-war years was under way . . . few cities were to reverse their planning policies as rapidly as Newcastle.' Burns was clearly more concerned with pursuing a vision of modernist planning than with the impact it may have had on displaced families. He commented about the break-up of communities that:

> One might argue that it is a good thing when we are dealing with people who have no initiative or civic pride. The task, surely, is to break up such groupings, even though the people seem to be satisfied with their miserable environment and seem to enjoy an extrovert social life in their locality.

A rather more sensitive approach was evident in the redevelopment of Byker in the 1970s by the British architect Ralph Erskine who worked for most of his life in Sweden. Interestingly, Smith argued that '[while] the opportunity to make something

of the spectacular topography of Newcastle, with its sweep southwards to the river, was lost . . . , a second chance . . . lies in the rebuilding of Byker'. The redevelopment comprised 1,800 homes built to replace the area's numerous terraces of Tyneside flats. Byker's housing had been described even in 1889 as 'unsightly bastilles of labour with barrack-like rows of dingy houses'. Although the replacement housing was at a relatively high density, the contrast in layout is spectacularly evident from comparing the two maps of Byker. Erskine aimed to rehouse the displaced residents so as not to break up social networks, although delays in the building process resulted in few of the original residents being rehoused in the new Byker. The most innovative element was the Byker Wall, an undulating block of flats over a mile in length and up to eight storeys in height. On its northern side it presented a virtually blank façade intended to insulate the flats from the noise of a proposed motorway. While the motorway was never built, the blank face offered protection against cold northern winds. Within the leeward side, low-level high-density housing was built, with different areas distinguished by varying designs and bright colours. Erskine also incorporated small private gardens and semi-public communal spaces. Drawing on his Swedish experience, he made great efforts to involve local residents in influencing the design, and the estate is seen as an example of the benefits of public participation in planning. Erskine's estate continues to be strongly lauded, and in 2007 English Heritage awarded it Grade II listing in recognition of its design and uniqueness.

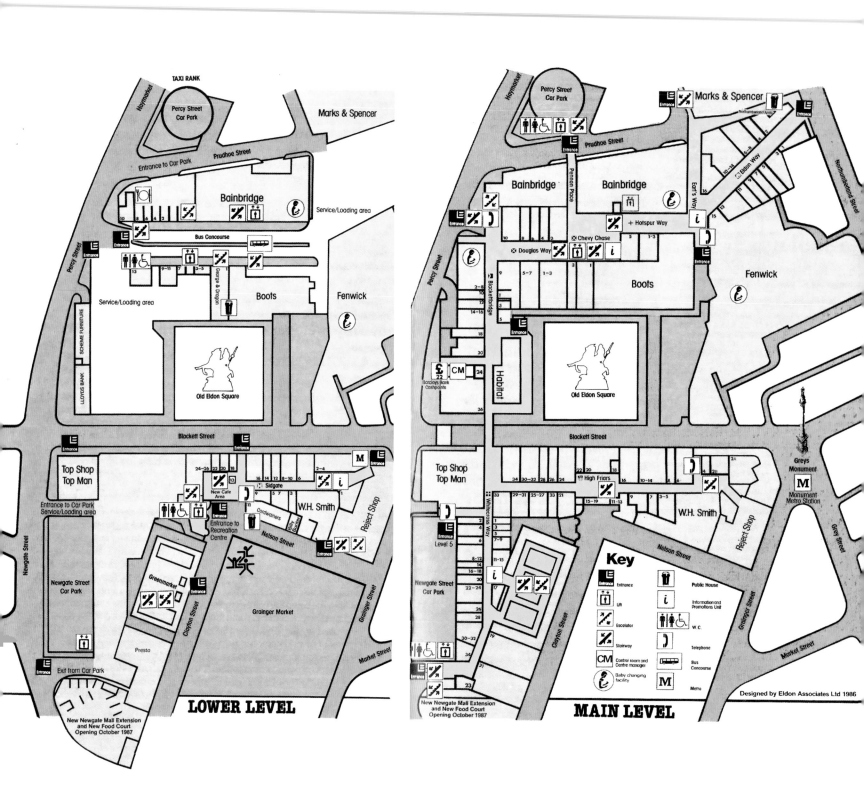

LOWER LEVEL

MAIN LEVEL

TAXI RANK

Percy Street
Car Park

Marks & Spencer

Entrance to Car Park

Prudhoe Street

Bainbridge

Service/Loading area

Bus Concourse

Percy Street

Entrance

George & Dragon

Boots

Fenwick

Service/Loading area

SCHEME FURNITURE

LLOYDS BANK

Old Eldon Square

Blackett Street

Entrance

Entrance

Top Shop
Top Man

New Cafe
Area

Sidgate

W.H. Smith

M

Entrance

Entrance to Car Park
Service/Loading area

Cordwainers

Billy Bookor

Reject Shop

Entrance

Entrance to Recreation Centre

Nelson Street

Newgate
Street

Newgate Street
Car Park

Greenmarket

Entrance

Grainger Market

Clayton Street

Grainger Street

Presto

Market Street

Exit from Car Park

New Newgate Mall Extension
and New Food Court
Opening October 1987

Percy Street
Car Park

Marks & Spencer

Entrance

Prudhoe Street

Northumberland Arms

Entrance

Bainbridge

Penton Place

Bainbridge

Hotspur Way

Earl's Way

Eldon Way

i

Entrance

Chevy Chase

Douglas Way

Blackettbridge

Boots

Entrance

Fenwick

Percy Street

Habitat

Old Eldon Square

Barclays Bank
Cashpoints

£

CM

24

Greys
Monument

Blackett Street

Top Shop
Top Man

High Friars

W.H. Smith

M

Monument
Metro Station

Whitecross Way

Nelson Street

Reject Shop

Entrance
Level 5

Newgate Street
Car Park

Clayton Street

Grainger Street

Grey Street

Market Street

Entrance

New Newgate Mall Extension
and New Food Court
Opening October 1987

Key

	Entrance		Public House
	Lift	i	Information and Promotions Unit
	Escalator		W.C.
	Stairway		Telephone
CM	Control room and Centre manager		Bus Concourse
	Baby changing facility	M	Metro

Designed by Eldon Associates Ltd 1986

1986

Shopping competition: Eldon Square versus MetroCentre

Retailing has undergone enormous changes in Tyneside and is . . . a dynamic and successful part of the local economy in spite of economic recession and high unemployment.

F. Robinson, *Post-Industrial Tyneside*, 1988

Tyneside is a classic consumer centre because of its relatively high disposable income that reflects the combination of salaries often set by national pay bargaining and living costs that are low relative to elsewhere in the country. The area has seen pioneering steps in retail provision, not least developing some of the earliest department stores.

Best known today are the Eldon Square Shopping Centre in Newcastle and Gateshead's MetroCentre, which at the time of opening were, respectively, the biggest covered shopping centre in the UK and the largest out-of-town shopping centre in Europe. These schemes brought competition to new heights, both in terms of rivalry between them and their impacts on the conurbation's retail geography more broadly. The tale of the two centres began in the 1970s, with the first stage of Eldon Square opening in 1976, followed ten years later by the Metro-Centre, and the rivalry continues to this day despite being owned by the same company in recent years – until it collapsed towards the end of the first Covid-19 lockdown in June 2020.

Their origins, however, are very different, as is their range of services. The Eldon Square Centre was built in the heart of Newcastle partly on the site of 'Old' Eldon Square. Part of T. Dan Smith's 'juggernaut of redevelopment', it involved the controversial demolition of all but the short eastern frontage of the Square's attractive terraced housing. It is readily accessible by public transport, notably by the Metro's Monument station which opened in 1980 and has direct connections to

---

OPPOSITE: Ten Years of Eldon Square Brochure, *Internal layout of Eldon Square* (1986) [NTCL] The plans
and design of the Eldon Square development evolved through no less than 32 schemes between 1968 and 1972.

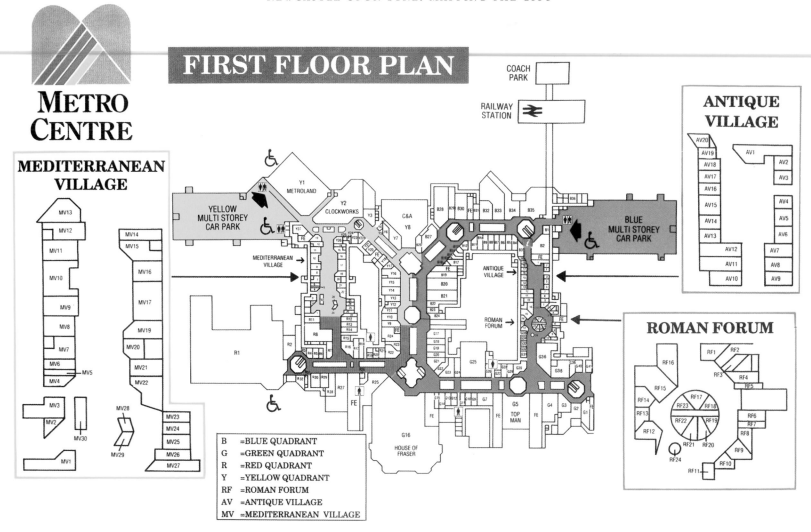

FIRST FLOOR PLAN

METRO CENTRE

MEDITERRANEAN VILLAGE

ANTIQUE VILLAGE

ROMAN FORUM

COACH PARK

RAILWAY STATION

YELLOW MULTI STOREY CAR PARK

BLUE MULTI STOREY CAR PARK

MEDITERRANEAN VILLAGE

ANTIQUE VILLAGE

ROMAN FORUM

HOUSE OF FRASER

TOP MAN

B	=BLUE QUADRANT
G	=GREEN QUADRANT
R	=RED QUADRANT
Y	=YELLOW QUADRANT
RF	=ROMAN FORUM
AV	=ANTIQUE VILLAGE
MV	=MEDITERRANEAN VILLAGE

MetroCentre Guide, *Internal layout of MetroCentre* (1986) [GLH] The retail development specialists Sovereign Centros took control of the Metro-Centre after the collapse of Intu, promising to inject new life into the Centre and make £25 million available for investment.

all points of the system, now including Sunderland. The principal anchor firm was Bainbridge's, a long-established local firm that is now part of the John Lewis Partnership, but the adjacent Fenwick's and Marks and Spencer stores facing onto Northumberland Street were linked into the development. While primarily a shopping centre (with over 80 outlets at the time of the Location Plan of 1986), it also included a leisure centre on an upper floor.

By contrast, the MetroCentre was built on a cleared power-

station site three miles west of Gateshead's town centre. Its construction was financed by the Church of England Commissioners and planned by Cameron Hall Developments. A major reclamation operation was facilitated by an Urban Development Grant and it became part of the Tyne and Wear Enterprise Zone in 1981, thereby benefitting from a ten-year property tax holiday and reduced planning regulation. The first two stages opened in spring and autumn 1986, followed by a third stage in October 1987. Now it boasts more than 370 shops with

190,000 m² of retail floor space, making it the second-largest shopping centre in the UK after London's Westfield.

Initially the MetroCentre's trading results were disappointing despite attracting Marks and Spencer's first out-of-town store as its anchor, located on the eastern side of the complex. Access was poor in its early days. Despite its name, it was not served by the Metro, and while a new station was opened on the Newcastle–Carlisle railway line and there was a network of bus routes on the south side of the Tyne, such services were relatively low-frequency. Initially, road access was also poor but this was greatly improved in the early 1990s by the completion of the Western Bypass and the building of a dedicated interchange, though this is still frequently congested at peak times. Public transport has been improved, notably with the opening of an integrated bus and rail station in the mid-2000s.

To increase its attractiveness, the MetroCentre diversified by incorporating a range of novelty features. These included three 'villages' with pseudo-heritage 'architecture' (the Mediterranean Village, the Roman Forum and the Antique Village), as shown on the First Floor Plan. Additional features were the Metroland funfair adjacent to the large car park at the Yellow Quadrant, a roller coaster running through part of the shopping centre, walkways broken up by spaces with giant chess boards, and large fish tanks that were intended to provide a tranquil, relaxing atmosphere. Not for nothing did it bill itself as the 'complete shopping experience'. While all of these features have since been removed, the combination of shopping and leisure remains, with the provision of a 12-screen cinema and numerous food outlets at the MetroCentre Qube. The Namco Funcentre provides family entertainment with dodgem cars and an 18-lane tenpin bowling alley. However, the development of a fifth 'mall', the Platinum, at a cost of £120 million, showed a stronger commitment to high-end retail provision.

Initially, it was thought that the MetroCentre would have a serious negative impact on Newcastle city centre, especially given the city's limited car parking with daytime charges in contrast to the 8,000-odd free spaces in the main MetroCentre complex (some 13,000 if including adjacent retail sites). But its effect over the first five years of trading was relatively small, a drop of no more than 8 per cent. Partly this was due to a nationwide surge in debt-fuelled retail sales in the 1980s, but it was also thanks to Eldon Square responding to the challenge. The latter's shopping malls were given their first major revamp then and its opening hours were extended, while, following the demolition of the Green Market and adjacent properties, it expanded southwards with the new St Andrew's Way anchored by Debenhams and providing 'high-end' clothing shops. A new bus station was provided and a further refurbishment took place in 2013. The new dining area, Grey's Quarter, provides a wide range of international cuisine. Eldon Square now provides sites for over 150 retail units.

The two centres also serve rather different markets. Geographically, the MetroCentre's catchment area is much wider, with over three million people living within two hours' drive and also, via the North Shields Ferry Terminal, attracting Scandinavian visitors, whereas Eldon Square relied mainly on customers from the central, northern and eastern parts of the conurbation. Demographically, the MetroCentre's clientele was drawn disproportionately from the ABC1 groups and the 25–34s, whereas Eldon Square exerted a stronger pull on less wealthy and older customers. MetroCentre's most serious impact was on the two local centres of Blaydon and Whickham as well as on central Gateshead itself, but none of these had been doing particularly well previously. Similarly, Eldon Square was found to have caused problems for some of Newcastle's neighbourhood shopping centres by capturing a larger share of the market, but its biggest negative impact was in streets immediately around it, especially Clayton and Grainger Streets. This followed the switch of the city centre's main access point from Central Station to the Metro's Monument station and was a factor in prompting the creation of the Grainger Town Partnership to revive the area.

Clearly, competition between shopping areas is still alive and well on Tyneside, but retailing appears to have entered a time of big uncertainty, not just with the collapse of the owner of the two centres but, potentially much more significantly, with the rising popularity of online shopping during the Covid-19 pandemic.

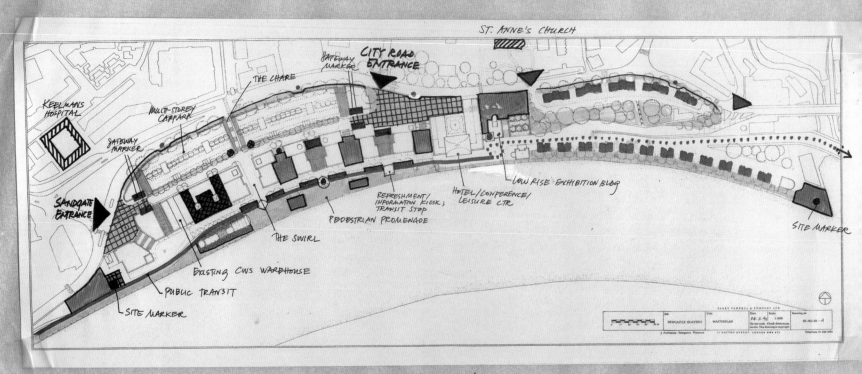

ST. ANNE'S CHURCH

KEELMANS HOSPITAL

MULTI-STOREY CARPARK

THE CHARE

GATEWAY MARKER

CITY ROAD ENTRANCE

GATEWAY MARKER

LOW RISE EXHIBITION BLDG

GATEWAY MARKER

SITE MARKER

SANDGATE ENTRANCE

REFRESHMENT/ INFORMATION KIOSK; TRANSIT STOP

HOTEL/CONFERENCE/ LEISURE CTR

PEDESTRIAN PROMENADE

THE SWIRL

EXISTING CWS WAREHOUSE

PUBLIC TRANSIT

SITE MARKER

Job: NEWCASTLE QUAYSIDE Title: MASTERPLAN Date: 24.2.96 Scale: 1:1000

SITE ACCESS

▶ ACCESS TO COMMERCIAL SECTOR

▶ ACCESS TO RESIDENTIAL SECTOR

⇢ EXIT ONLY (RESTRICTED ENTRY)

⌣•⌣ BUS STOP LAYBY

▭ LANDSCAPE SCREENING

MASTERPLAN

▭ PEDESTRIAN PROMENADE

▬ GUIDED MODE TRANSIT

•••• MANUAL MODE TRANSIT

▬ VIP/ DISABLED PARKING

▭ DISABLED ACCESS   ⊗ LIFT

▬ RESIDENTIAL UNITS

▬ URBAN SQUARES

▨ MAIN URBAN SQUARE

▬ SITE MARKER

▨ GATEWAY ENTRANCE OPEN SPA

▬ GATEWAY MARKERS

1990

Regenerating the Newcastle–Gateshead Quayside

Like the river itself, the Newcastle–Gateshead Quayside never stands still. It has become a spectacular place to see and experience, yet only a few decades ago the look, the atmosphere and even the reputation of the area were quite different.

I. Ayris, *Unlocking the Quayside*, 2006

Even long before the compulsory introduction of town planning by every local authority in 1947, the map had been seen as an essential planning tool. Planning maps enable decision makers and the public to envisage the scale and character of proposed developments and assess their relationship to existing environments. In late-twentieth-century Britain, with the burgeoning number of redevelopment schemes whose focus is primarily on property-led regeneration, the map has increasingly taken on a less objective role. Along with illustra-tions such as the 'artist's impression', the map has often become part of a marketing package, the purpose of which is to 'sell' the scheme under consideration. Such 'promotional' represen-tations are, in many ways, the modern equivalent of medieval picture-maps.

A key instrument in the delivery of such regeneration schemes was a series of development corporations, established by the Conservative government of the 1980s. These appointed bodies took control of specific designated areas away from elected local authorities and consisted mainly of local business interests. The Tyne and Wear Development Corporation (TWDC) was established in 1987 for a ten-year period, tasked with fast-tracking economic, social and environmental devel-opment in areas that had become largely derelict. It had extraordinary powers to compulsorily acquire land and functioned as the statutory planning authority for its areas

OPPOSITE: Sir Terry Farrell, *East Quayside Masterplan sketch* (1990) [ROB]
Sourced from the Terry Farrell Archive recently donated to the Robinson Library.

215

during its lifetime. The TWDC was itself the culmination of a series of inner-city policy initiatives. For example, under the 1978 Inner Urban Areas Act, the riverside areas of Newcastle and Gateshead had been chosen as one of seven participating areas in an Inner-City Partnership Scheme. This had already led to major regeneration proposals for the Newcastle–Gateshead Quayside, published in the *Framework for Action* document of 1981. Ideas and initiatives to regenerate the Quayside area were not new, the powers to do so were.

The TWDC's flagship plan, Newcastle East Quayside, was one of three major regeneration areas along the Tyne (the other two were Newcastle Business Park on the site of Armstrong's former Elswick Works upstream and the Royal Quays retail park and housing development in North Shields, built around the former Albert Edward Dock). The 25-acre East Quayside area had formerly been at the heart of Newcastle's economic activity, but during the twentieth century the focus of the latter shifted dramatically to other locations, leaving the Quayside effectively redundant.

The main map shows one of a series of preparatory sketches drawn by consultants Terry Farrell & Partners for Newcastle's East Quayside. Others included landscaping and planting features. The scheme envisaged a sequence of landscaped squares and privileged pedestrian access. A major square was planned for the heart of the area linking commercial activity, parking and pedestrian movement. Development was scheduled to proceed from east to west in a series of phases. In fact, this was the second set of proposals for the area, as an earlier scheme by Rosehaugh/Shearwater had collapsed when that firm experienced financial difficulties. Their plan had featured significant office development and so-called 'festival' shopping, both opposed by Newcastle City Council due to their concern about impact on the city centre. Farrell's plan was more varied and sought to develop a mix of commercial, residential and leisure facilities including a 150-bed luxury hotel providing conference amenities, and new pubs and restaurants fronting the south-facing promenade along the river. The decision to refit the Grade II listed, former CWS warehouse as a Malmaison Hotel has been successful. Although some of the

details of the scheme had to be adjusted (the proposed hotel and conference centre did not attract developers, for example), the area has emerged broadly in accordance with the plan. The eastern part of the area has successfully attracted residential apartment development, although in March 2021 councillors rejected a 14-storey 289-apartment block on the last remaining site at St Ann's Quay as inappropriate in scale and undistinguished in design.

While the TWDC scheme has undoubtedly been successful in changing the physical appearance of the Quayside, a number of other factors have impacted powerfully on the area. For example, the publicly funded new Law Courts (1984–90) built immediately to the west of the TWDC site was a catalyst for the subsequent overall regeneration of the Quayside. It attracted legal firms to the area, including Newcastle's two largest firms of solicitors, Dickinson Dees (which relocated from the Westgate Road area) and Ward Hadaway. Other legal firms followed suit, along with several property firms. It may be no coincidence that high-quality restaurants have subsequently been one of the major leisure functions to locate within the area.

Although not part of Farrell's remit, linking the two banks of the river is the Gateshead Millennium Bridge which has become a major attraction in its own right. It was funded by the European Regional Development Fund (ERDF) and the Millennium Commission as part of Gateshead's imaginative regeneration programme. In contrast, Gateshead Quayside regeneration has had to proceed piecemeal as the Council has been unable to afford the purchase of the huge amount of land required for a comprehensive and fully integrated scheme. However, Gateshead already had an impressive track record of cultural-led initiatives (the most publicised being the Angel of the North statue and the 1990 Garden Festival attracting more than 3 million visitors) and, against this background, the Council sought to develop the 'Gateshead Quays' area. Baltic Quay, originally a housing development of 241 apartments, has recently moved into a second phase, while development is also proceeding on Baltic Business Park, costing £250 million and containing a 'Knowledge Campus' based on the relocated Gateshead College. The former Baltic Flour Mill (built in 1950

Gateshead Quays Development Image (2018) [CONFERENCE NEWS MAGAZINE] One of several images produced by the consortium of Gateshead Council, Ask Real Estate, PATRIZIA and architects HOK.

on the site of Hawks and Crawshay's Gateshead Iron Works which smelted the iron for the High Level Bridge) has been imaginatively converted as the Public Centre for Contemporary Art (the Baltic Centre), partly using Lottery funding to support culture and the arts. Even more spectacular is the Sage Music Centre designed by Norman Foster, a shell-like structure of stainless steel and glass containing three auditoria, the largest holding 1,700 people. It provides a world-class education and performance centre for every type of music. Although criticised by some for being out of scale with its environment, the Sage certainly makes a dramatic addition to what was already a spectacular visual environment.

More recent proposals for the Gateshead Quayside have, however, attracted more criticism. An 'artist's impression' of the Gateshead Quays proposal for the area in between the Sage and the Baltic buildings is shown above. This £260 million proposal by HOK, specialists in arena development, in conjunction with AHR architects won planning approval from Gateshead Council in November 2020. Covering ten acres overall the focus of the development will be the Arena performance venue, to be relocated here by 2023 from Forth Banks in Newcastle, with a capacity of 12,500 people. Also included in the complex is a conference centre (perhaps it will have more

luck than the one proposed by Terry Farrell Partners across the river), three large exhibition spaces, external performance spaces, a 300-room hotel, shops, restaurants and a range of public spaces. It is claimed the development will create 2,000 new jobs and bring in £30 million to the local economy annually. Despite the impressive statistics, public reaction has been less than enthusiastic with one architectural critic describing the proposal as 'a crumpled mishmash of competing structures', and a key point is that there appears to be confusion between attempting to develop another landmark building (to go with the Baltic and the Sage) and simply filling in the available space between the two. There is a genuine sense of disappointment at what has now been approved for one of the keynote sites in the north of England.

Furthermore, the proposed development puts one of the key questions about property-led regeneration into sharp focus, namely who is it really for? The contractor for the site is the ubiquitous Sir Robert McAlpine, well practised in the business of profitable but tediously convergent redevelopment of urban areas. Meanwhile, in Gateshead, 32,700 people live in the most deprived 10 per cent of localities in England. There is little surprise if they regard this development as having nothing to do with them.

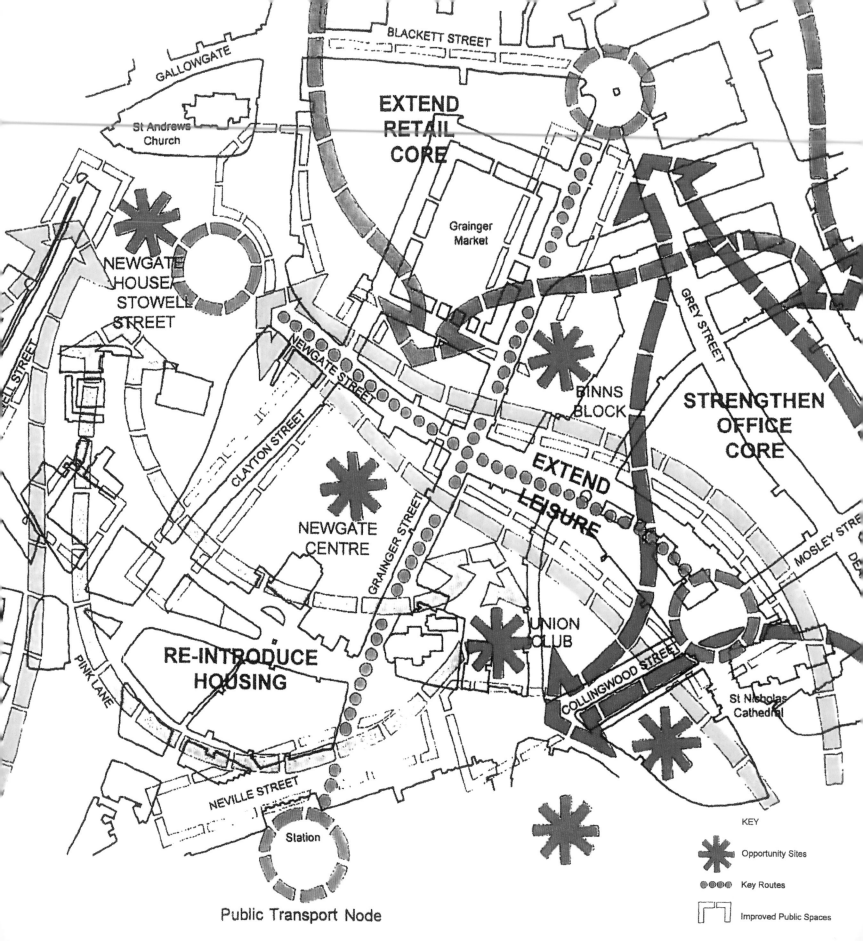

EXTEND
RETAIL
CORE

GALLOWGATE

BLACKETT STREET

St Andrews
Church

Grainger
Market

NEWGATE
HOUSE/
STOWELL
STREET

GREY STREET

STRENGTHEN
OFFICE
CORE

NEWGATE STREET

BINNS
BLOCK

CLAYTON STREET

EXTEND
LEISURE

NEWGATE
CENTRE

GRAINGER STREET

MOSLEY STREET

PINK LANE

RE-INTRODUCE
HOUSING

UNION
CLUB

COLLINGWOOD STREET

St Nicholas
Cathedral

NEVILLE STREET

Station

KEY

Opportunity Sites

Key Routes

Public Transport Node

Improved Public Spaces

1996

Rescuing Grainger's legacy

The name Grainger Town was first applied to this city quarter in 1991 to give a distinctive and memorable identity to what was then an embryonic revitalisation project: this has since blossomed into the Grainger Town Regeneration Project as we now know it.

D. Lovie, *The Buildings of Grainger Town*, 1997

Central Newcastle has been called 'a City of Palaces'. Shortly after its completion, Gladstone described Grey Street as 'Britain's best modern street', and its graceful curve was highly praised by John Betjeman in 1948. By the 1980s, however, the object of this admiration – the group of streets developed by Richard Grainger and his associates in the 1830s – was in a sorry condition. A survey of 1992 found that, despite one-third of its 420 buildings being either Grade I or Grade II listed, over half were deemed to be 'at risk', while more than one-third of their total floor space was totally unused. At the same time in 1992, over 800 commercial and cultural businesses remained in the area, understandably becoming increasingly anxious about the long-term future of these streets.

This was the context for the establishment of the six-year (1997–2003) Grainger Town Project and the catalyst for a notable transformation in the fortunes of this key part of Newcastle's heritage. Winning a British Urban Regeneration Association Best Practice award as early as 2001 and commendation from the Royal Town Planning Institute, it is now widely recognised as an exemplary regeneration scheme and its approach has been applied both elsewhere in the UK and abroad.

The Project focused primarily on the 1830s–40s Grainger developments but, as the strategy map shows, also extended westwards to the city wall to include the Blackfriars area and

---

OPPOSITE: Grainger Town Project, *Grainger Town Regeneration Strategy* (1996) [NERA] The area developed by Richard Grainger constitutes only about 25 per cent of the overall Grainger Town Project area.

Newcastle's 'Chinatown' in Stowell Street. The whole area had been in decline since the 1950s, when Grainger Street's mantle as the city's premier shopping street passed to Northumberland Street. This northward shift in the focus of shopping activity was signalled by the loss of several department stores, including Wengers, Farnons and Binns, and was reinforced in the 1970s by the building of the covered Eldon Square shopping centre and the opening of the Metro transit system under the city centre. Then in the 1980s, much business activity was lured away to the new offices built as part of the regeneration of the Quayside and the opening of Newcastle Business Park on the site of Armstrong's former Elswick works.

The Grainger Town Project had its origins in 1991 when Newcastle City Council and English Heritage jointly commissioned a ten-month study of the area. This resulted in the Council being invited to join a pilot Conservation Area Partnership, one of 15 such pilots set up around England to explore an entirely new and innovative way of delivering grant aid. For the one-year pilot in 1994/95, the Partnership prepared an Action Plan and Bid for £300,000 which, along with City Council funding, would be used to help rescue historic buildings in critical condition, demonstrate what was possible in terms of repairing problem buildings and reusing vacant upper floors, raise the quality of street façades and promote further action via education and information.

This in turn led to the appointment of consultants in 1996, tasked with drawing up a comprehensive strategy for regeneration. The resulting project was an ambitious £120 million heritage regeneration programme which would lead to Grainger Town becoming 'a dynamic and competitive location in the heart of the city', with a focus on leisure, culture, the arts and entrepreneurial activity, and making it 'a safe and attractive location to work, live and visit'. Improving the competitiveness of the local economy was central to its strategic objectives, recognising that the regeneration of the area was not simply about architectural heritage. The accompanying map shows the overall strategy. Key 'opportunity sites' were identified where often long-empty property, such as the Union Club in Westgate Road, could be brought into use. Recognising

the functional diversity of the modern city centre, the strategy sought to develop two leisure 'axes', one in the west extending from the Blenheim Street area with its clubs and dance studios northwards to Stowell Street (Chinatown), and one in the east to build on the already lively Bigg Market area but extending this to include large projects such as the Gate leisure complex in Newgate Street. The strategy also sought the strengthening of office provision, especially in the western part of the area, and the recolonisation of Grainger Street and Clayton Street, in particular by retail functions. The vacancy map of 1996 shows that a major problem existed with unused upper floorspace, particularly in Grainger Street and the west side of Clayton Street. With this amount of under-used space in upper floors in mind, the encouragement of residential accommodation 'above the shop' was a core theme of the strategy, along with conversions for larger residential units including the long-awaited demolition and redevelopment of the widely detested Westgate House building – this for student accommodation.

The Grainger Town Project's achievements have been impressive. An estimated public-sector investment of £40 million led to private sector investment of £145 million. By 2003, 1,500 jobs had been created directly by the project, with a further 800 jobs in Grainger Town generally. Almost 300 new businesses had been set up and, also reflecting increased confidence in the area, retail and office rentals rose by 60 and 38 per cent respectively between 1997 and 2001. Despite the economic downturn of the early 2000s, progress continued. The number of residential units increased from 998 in 2003 to 1,455 in 2009, many within buildings previously classified as being 'at risk'. The population of the project area was calculated to be around 800 persons in 1996 but was well over 2,000 by 2009. Significant doubts had been expressed about the existence of a market for city centre residential developments, but this success demonstrates that these were ill-founded. Between 1997 and 2000, 116 residential units were created in the Grainger Street and Clayton Street areas with the help of the 'Living Over the Shops' scheme, funded by the Single Regeneration Budget. Particularly significant was the creation of 23 residential units in the upper floors of the Central

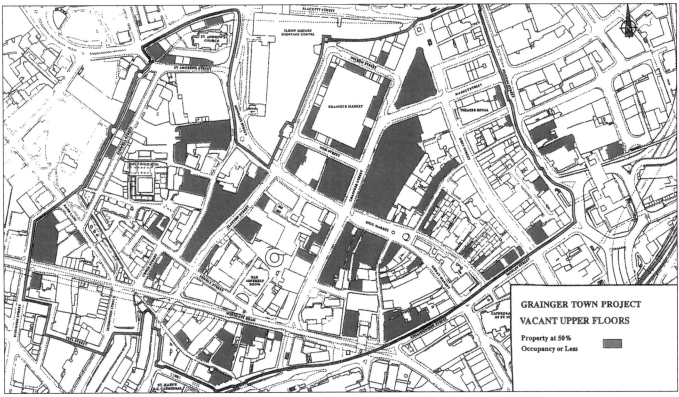

GRAINGER TOWN PROJECT
VACANT UPPER FLOORS

Property at 50%
Occupancy or Less

Grainger Town Project, *Grainger Town Vacant Upper Floors* (1996) [NERA] The unused floor space in
Grainger Town was sufficient to house 7,000 people or workspace for 11,000 office workers.

Exchange Building housing the iconic Central Arcade. This development was carried out entirely without public funding, and by the early 2000s all the units, of varying sizes, had been sold for prices ranging from £220,000 to £560,000.

Through the first decade of the 2000s the number of business premises in the Grainger Town area increased by 9 per cent, from 939 to 1,024. Above all, there has been significant improvement to the physical fabric of the area with a 2010 survey assessing 97.7 per cent of buildings to be in 'good' or 'fair' condition. Improvements to the public realm include pavement renewal, shopfront design initiatives and the creation of a sense of unity through street name-plate design.

In addition, many business, arts and cultural events were organised and supported, including an international conference which focused on the regeneration of historic cities and launched a network of cities dedicated to the promotion of best-practice in heritage-led regeneration. Key lessons learned from the project included the value of a heritage audit and detailed survey at the start of the programme, the need to change perceptions of heritage by promoting a distinctive civic vision and the importance of 'early wins' based on the selective use of compulsory purchase orders to demonstrate resolve. Also vital in schemes such as this is the preparation of a forward strategy, which the Grainger Town Partnership prepared in line with Newcastle City Council's Unitary Development Plan during the last two years of the project. Above all, the project has secured the long-term future of the area's impressive architectural heritage, confirming Newcastle's place as an internationally significant historic centre.

1999

The Great North Run

The 'Great North Run' has been the most obvious manifestation of that need to participate and to identify with the top performers.

H. Taylor, *Geordies*, 2005

This image of the 1999 Great North Run is by the Byker-born painter, John Coatsworth. His characteristic 'curvation' style is appropriately employed here in displaying a unique event. Many cities have marathons and half-marathons but the Tyneside event is distinctive in several respects. Most urban centres where such events take place are instantly recognised iconic, global cities, which Newcastle cannot really claim to be, despite its role as regional capital of the North East. But its Great North Run is the largest half-marathon in the world with over 57,000 participants. It also retains its distinctive atmosphere of an elite international athletic competition

sharing the same course with 'fun runners', many dressed in bizarre costumes. It is one of the few events in the world where the cream of middle- and long-distance runners participate at the same time as enthusiastic amateurs of all ages and capabilities. Its egalitarian character is indicated by the fact that the millionth runner over the finish line was a female fun runner, Tracey Cramond, in 2014. No other similar event in the world has recorded anything like this number of participants. Coatsworth's image captures the locally grounded nature of the event, its intensity and especially its fun element.

The first Great North Run took place in June 1981 and with 12,000 people taking part, it clearly had the potential to become something more substantial. This was largely due to the popularity of the organiser, the Hebburn-born middle-distance runner Brendan Foster. Foster had become a national sporting hero in the mid-1970s, with a stunning series of

---

OPPOSITE: John Coatsworth, *Cartoon of BUPA Great North Run: Crossing Tyne Bridge 1999* (1999)
[JOHN COATSWORTH] Coatsworth's distinctive style captures Newcastle scenes with colour and humour.

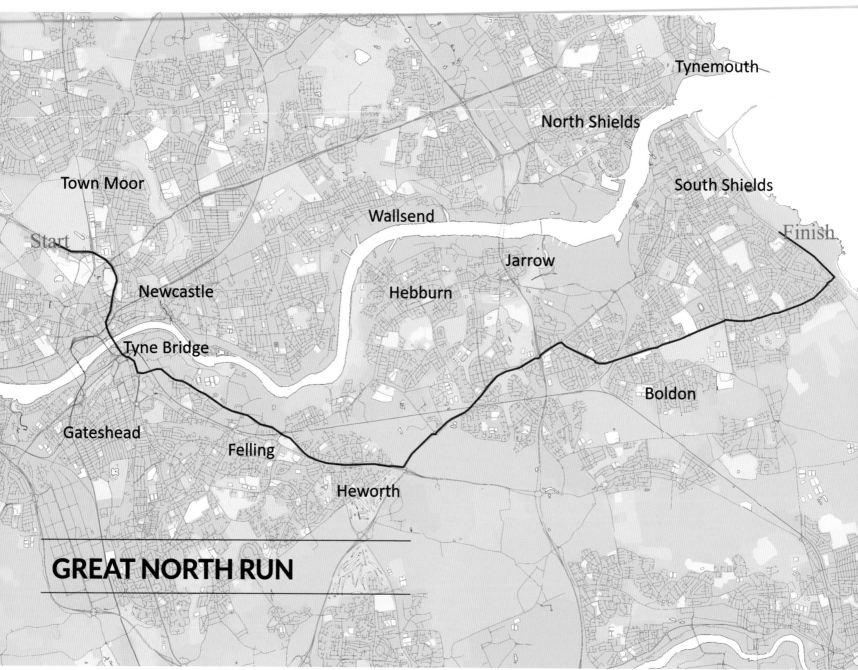

Sam Atkinson, *Route of Great North Run* (2020) [LOVE TO RUN CO.]
The course is relatively flat with the highest point being 60 metres at
the Black Bull junction, Heworth, and the lowest ten metres at the
John Reid Road roundabout in the southern part of Jarrow.

record-breaking victories. Having broken the world record for two miles at the Crystal Palace in 1973, in 1974 he broke the 3,000 metres world record and followed this by winning the European Championship 5,000 metres race in Rome in a record time. Despite illness, he won the bronze medal at the 1976 Montreal Olympics 10,000 metres race – the only British track and field athlete to actually win a medal at those games. If these achievements made him a national sporting celebrity, he was even more of a local hero due to his approachable personality and embeddedness within the region. He taught chemistry in his home town at St Joseph's Grammar School, Hebburn, where he had been a pupil – a fact which largely explains the route of the Great North Run.

Brendan Foster had organised a local 'Gateshead Fun Run' in 1977 but was inspired to initiate the more ambitious Great North Run through his own experience of the New Zealand 'Round the Bays' race in 1979 where fun runners mixed with serious athletes. Conscious that the North East's sporting image was dominated by association football and participation in athletics was very low, Foster sought to popularise athletics in a different way, using his fame to develop what has become a leading event in the international sporting calendar. The winner of the first Great North Run in 1981 was the Northumbrian Mike McLeod in a time of one hour and three minutes. The competitiveness of that first event is signalled by the fact that 40 years on, only four minutes have been knocked off that time, the record now being held by the Kenyan Martin Mathathi at 58 minutes 56 seconds. A fellow Kenyan athlete, Mary Keitany, holds the women's record of one hour five minutes and 39 seconds. Mo Farah won the event five times in succession from 2014 to 2018.

The 13.1 miles of the run is unashamedly urban, starting from just north of Newcastle's city centre and finishing at the coast in South Shields. Each year around a quarter of a million spectators line the route through Gateshead and South Tyneside. Appropriately enough, given the site's historical role in communal gatherings, the run begins on the southern edge of the Town Moor and proceeds along the Central Motorway on the eastern side of the city centre to the Tyne Bridge. Over in Gateshead it swings to the east parallel to the Tyne and along the Felling bypass, then heading north-east along Leam Lane (A194) to the southern edges of Hebburn, Jarrow and South Shields before reaching the coast at Marsden. The last mile is northwards along the seafront to the finishing line at South Shields.

The Great North Run is always held on a Sunday, but the family-oriented nature of the event is indicated by holding the United Kingdom's biggest running event for children on the preceding day. Saturday morning sees a 'Mini Great North Run' over 1.5 kilometres for children aged three to eight years, and later in the day a 'Junior Great North Run' (4 kilometres) is held for nine to 16 year olds. The routes for them are on the regenerated Newcastle and Gateshead Quaysides. The 'Mini' run starts on the Newcastle side, crosses the Swing Bridge, turns east along the Gateshead Quayside before crossing back to the Newcastle side via the Millennium Bridge. The 'Junior' run follows a similar route but, after crossing the Swing Bridge, turns westwards along the south bank of the Tyne before doubling back to pass the Sage Music Centre and crossing the river on the Millennium Bridge.

These events have become a major 'booster' to the region's economy and image, and the economic statistics are impressive. In 2017 over £18.6 million was raised for charities and the event itself generated an estimated £23.7 million for the regional economy. At the 2017 run, participants came from 178 different countries. A survey in 2019 calculated the regional income generated by the event had increased to £31 million. Over 60 per cent of the runners travel to the event from outside the North East and most bring at least one other person with them. The average per capita expenditure for those staying overnight in the Newcastle area was over £500. About two-thirds of the non-local participants said that their view of the region had changed positively after taking part in the competition. The Great North Run has become famous for attracting celebrity participation and raising money for charitable causes – a tone that was set at the first Great North Run in 1981 by Newcastle United's Kevin Keegan who promised to donate 50p to charity for every person who beat him.

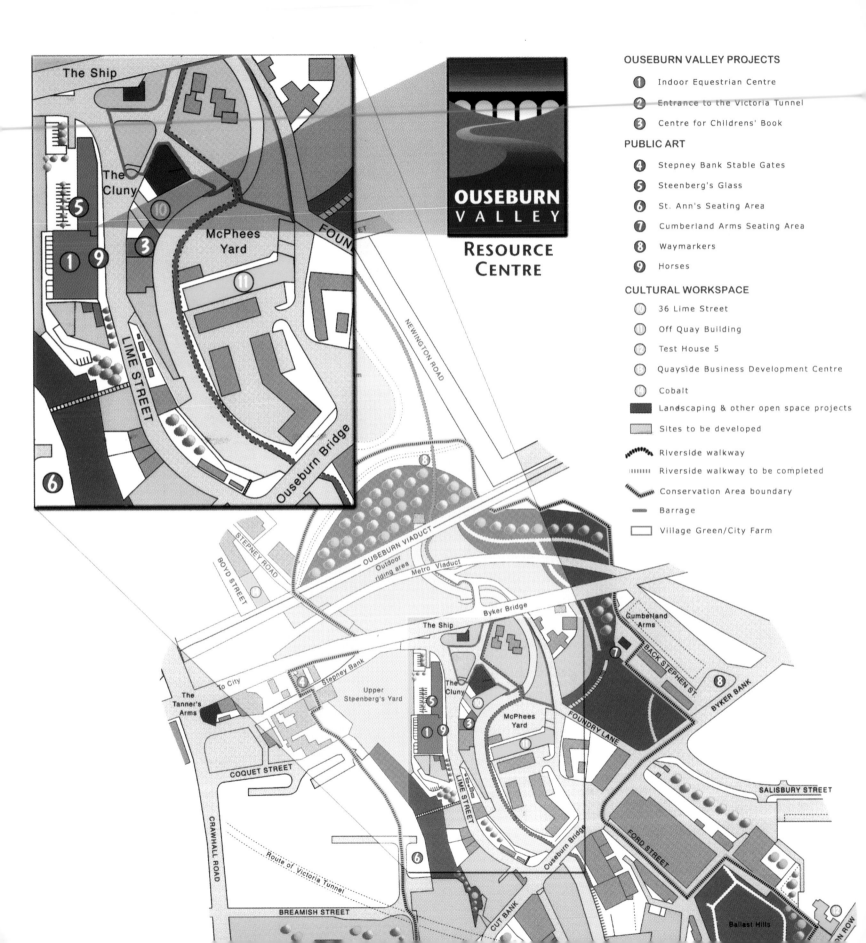

OUSEBURN VALLEY PROJECTS
1. Indoor Equestrian Centre
2. Entrance to the Victoria Tunnel
3. Centre for Childrens' Book

PUBLIC ART
4. Stepney Bank Stable Gates
5. Steenberg's Glass
6. St. Ann's Seating Area
7. Cumberland Arms Seating Area
8. Waymarkers
9. Horses

CULTURAL WORKSPACE
10. 36 Lime Street
11. Off Quay Building
12. Test House 5
13. Quayside Business Development Centre
14. Cobalt

Landscaping & other open space projects
Sites to be developed
Riverside walkway
Riverside walkway to be completed
Conservation Area boundary
Barrage
Village Green/City Farm

OUSEBURN VALLEY RESOURCE CENTRE

2003

A new role for the Ouseburn Valley

The Ouseburn would only a few years ago have seemed an unlikely site for . . . an urban village beside a clear stream, the sort of area in which people might stroll, sit outside pleasant pubs and visit art galleries and craft workshops.

A.W. Purdue, *Newcastle: The Biography*, 2011

The Ouseburn Valley lies immediately to the east of the TWDC's Quayside area and has been the location for a very different style of regeneration programme, one that is 'bottom-up' rather than 'top-down'. The map provides a good sense of the eclectic nature of activity in the area, a feature that adds considerably to its character as a distinctive place. Ouseburn was Newcastle's first major heavy industrial district with glass works, potteries, lead and iron factories. Their closure left an environmentally scarred area with many empty buildings, most in very poor condition, but with some occupied by small businesses taking advantage of relatively cheap rents. Although slum clearance had removed virtually the entire residential population, several pubs in the valley maintained a lively social scene.

In 1988 the 'East Quayside' group emerged from the intense activity of several bodies in the 1980s, with the main concern of ensuring that future development was carried out 'in a way which involves those who live and work in the valley or close to it'. Important constituents were the City Farm, opened in the valley in the 1980s, and an artists' cooperative based at 36 Lime Street Studios in the former Cluny Whisky warehouse (number 10 on the map). Two catalysts stimulated local activity. First, the TWDC East Quayside area with its property-led, large-scale business focus raised concerns that this approach would spill over into the Ouseburn Valley. The

OPPOSITE: Ouseburn Partnership, detail of *Ouseburn Valley Projects Plan* (2003) [OUSEBURN PARTNERSHIP]
The inset map shows the core area of the Ouseburn Regeneration projects.

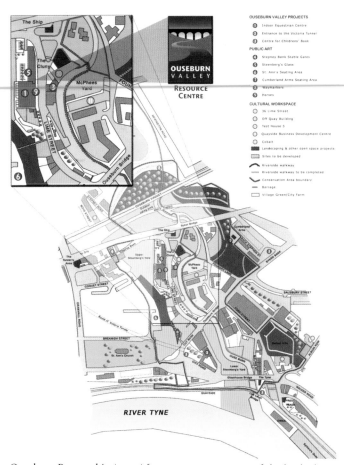

OUSEBURN VALLEY PROJECTS
1 Indoor Equestrian Centre
2 Entrance to the Victoria Tunnel
3 Centre for Childrens' Book
PUBLIC ART
4 Stepney Bank Stable Gates
5 Steenberg's Glass
6 St. Ann's Seating Area
7 Cumberland Arms Seating Area
8 Waymarkers
9 Horses
CULTURAL WORKSPACE
10 36 Lime Street
11 Off Quay Building
12 Test House 5
13 Quayside Business Development Centre
14 Cobalt
⬛ Landscaping & other open space projects
▢ Sites to be developed
〰 Riverside walkway
⋯⋯ Riverside walkway to be completed
— Conservation Area boundary
▬ Barrage
▢ Village Green/City Farm

OUSEBURN
VALLEY
RESOURCE
CENTRE

Ouseburn Partnership (2003) [OUSEBURN PARTNERSHIP] An intriguing
and compact mix of housing, industry, heritage, leisure and
entertainment.

second was a major fire at the empty Maynard's Toffee Factory
in 1993, raising the likelihood of the demolition of this
landmark building along with the loss of the other historic
buildings around it. A groundswell of opposition led to the
creation of the Ouseburn Trust in 1996 as a charitable organisation, seeking to develop the area on terms set by local
groups. A development partnership rapidly emerged, consisting
of the Trust and 18 other organisations, including Newcastle
City Council. This partnership functioned from 1997 to 2002
with a focus on 'grass roots', small-scale initiatives aimed
especially at creative industries. Significantly, Tyne Tees TV
had operated on the edge of the valley until 2005 and the area
benefitted from spin-offs from this proximity, particularly in
the music business. Number 11 on the map is the Off Quay
Building, providing cheap access to music studios and rehearsal

space for many years. Similar affordable space was created for
artists and designers. In 2003 a formal regeneration strategy
was published, with strategic responsibility vested in the
Ouseburn Trust which still maintained its charitable status.
Research for the regeneration strategy showed that there were
now 300 businesses in the valley. From 2004 most of the valley
was designated as a Conservation Area (boundaries shown on
the map) and, in April 2009, the Victoria Tunnel (a former
underground waggonway running 2.4 miles from Spital
Tongues and number 2 on the map) was reopened as a major
tourism and educational centre. The tunnel is now managed
as a visitor attraction by the Ouseburn Trust.

Between 2003 and 2011, £67 million was invested in the
valley, £25 million of it being public money. The strategy was
to build on and enhance what was becoming one of the region's
most dynamic creative and cultural clusters. Over 50 individual
physical regeneration projects were progressed, involving
mixed-funding conversions of existing buildings and some new-
build schemes. However, some subtle changes were taking place
in the nature of economic activity in the Ouseburn. There was
a notable growth in PR and marketing businesses along with
software, architecture and landscaping practices, printing and
publishing, as well as film and video production. The conversion of a former school into a Business Development Centre
(number 13 on the map) is illustrative of this changing
economic context. A significant indication of this second and
rather different phase of Ouseburn's 'revival' is indicated by
the former Maynard's Toffee Factory, which was rebuilt and
reopened in 2012, housing 25 digital businesses in the fields
of design, advertising and marketing.

Just as the Law Courts can be argued to have been the key
catalyst for the successful regeneration of the Quayside, the
location of Seven Stories (the National Centre for Children's
Books, number 3 on the map) in the heart of the Ouseburn
area in 2005 made a positive statement about the area's future.
It is the only centre for children's books in the UK and collects
archival material relating to the writing and illustration of this
literature. The Centre is a registered charity, located in a seven-
storey Grade II listed former warehouse built in the 1870s.

Funding was provided by the European Union (£1.24 million), Newcastle City Council (£1.4 million) and OneNorthEast (£1.4 million). The Centre includes a museum, two galleries, a book depository, an interactive discovery centre, an arts and education studio, digital facilities, a bookshop and a café. It attracts about 80,000 visitors per year, playing a significant role in the repositioning of the Ouseburn as a 'cultural' destination.

Not highlighted on the map is probably the most significant change occurring within the lifetime of the Ouseburn regeneration activity, namely the return of a residential population. A considerable level of interest has been shown in the area by housing developers, particularly since the millennium. Some regard this as a positive sign of the area's success. Others are less enthusiastic and fear an influx of gentrification. But, in fact, the first residential development was carried out by a housing association with the Tyne HA's renovation of Lower Steenberg's Warehouse and creation of 11 flats for vulnerable single people.

Local suspicion over housing developments was exacerbated in the early 2000s with a proposal by Wimpeys to build a massive 32-storey block of luxury flats at Malmo Quay where the Ouseburn meets the Tyne. Protests caused this to be withdrawn but it was soon replaced with a proposal to build a 13-storey structure. Planning permission was refused by Newcastle Council but, on appeal, was given the go-ahead by the government planning inspector. Though it has never been built, at the time of writing (March 2021), a new proposal – this time for an 18-storey tower block – has been submitted. The earliest large private scheme was a gated community with a residents' gym and on-site concierge at Lime Square on the western side of the valley in 2006, consisting of 104 one- or two-bedroomed apartments, three penthouses and four 'city houses'. A rather different clientele is served by another of the Tyne HA's developments, namely Farm View in Foundry Lane, 42 flats opened in 2015 and providing accommodation for vulnerable and isolated people. A still different component has been added to the residential mix with the creation of 266 student apartments at Stepney Yard and the building of four

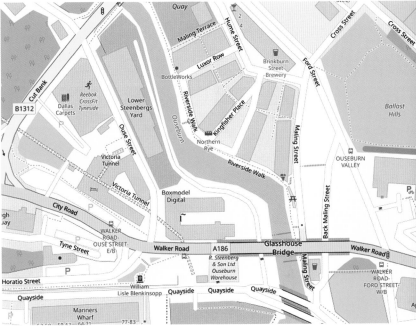

Open Street Map, *Lower Ouseburn* (2020) [OSM] The Malings development is bounded by Maling Street, Riverside Walk and Luxor Row. Boxmodel Digital is the site of the former Maynard's Toffee Factory.

blocks of student flats in Coquet Street, overlooking the valley.

An award-winning private residential development scheme is shown on the map above. The Malings is a high-quality development of 76 homes of varying sizes, designed with the creation of a coherent community as its core focus. Winner of the 2016 Housing Design Award, the scheme has won much praise. Streets run down the contours of the valley and the area is designed as a sequence of spaces of varying character. 'Traditional' terraced streets are mixed with courtyard dwellings and tower houses with garden space on their flat roofs. No two houses are identical and all have either an outdoor garden space or roof terrace. Communal facilities are arranged to encourage social mixing. Residents have clearly bought in to the concepts behind the development as, since completion, a combined pottery/café and a microbrewery/beer garden have opened. In February 2021 a new community garden proposed by local residents was given planning permission by the City Council.

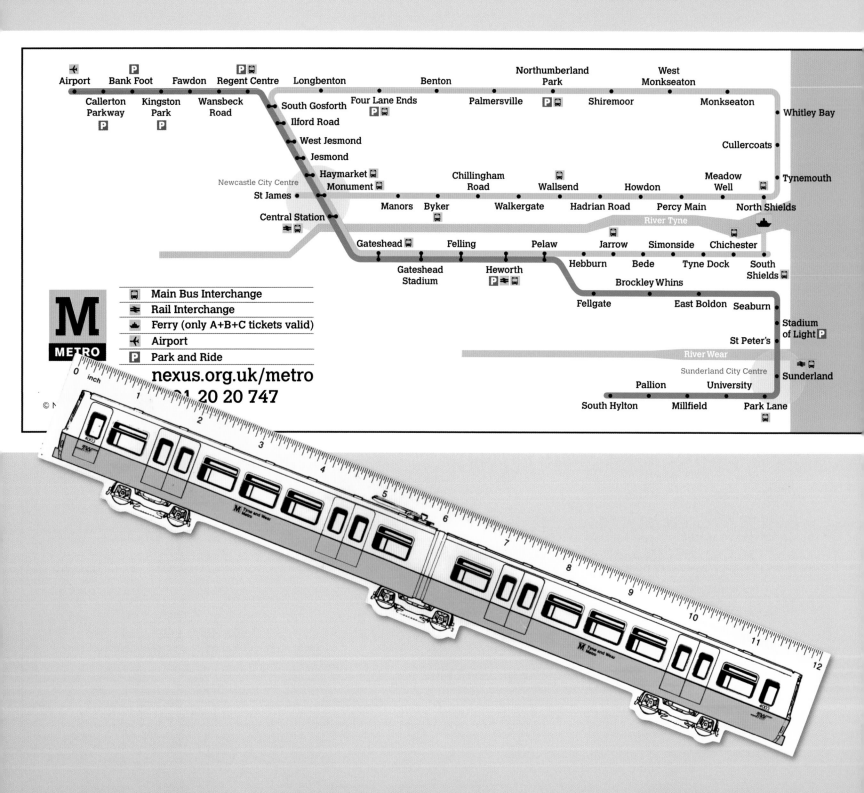

2008

The Tyne and Wear Metro

Metro is bound to tighten the links between the people
who live along the rest of the Tyne and Newcastle.

D. Bean, *Newcastle 900*, 1980

There may be questions about the role of the Metro in
promoting social integration on Tyneside, but there is little
doubt about its major impact on travel patterns for work,
shopping and leisure throughout much of the area. The Metro
is one of Tyneside's greatest post-war achievements, continuing
the strong local record of transport innovation associated with
the likes of Stephenson, Parsons and Merz. It was the first Light
Rail Transit (LRT) system in the UK, the first railway in the
country to be fully wheelchair-compatible, and one of the first
in the world to be completely non-smoking, as it was from its
inception (London Underground followed suit four years later).

It was the first example of the conversion of underused
suburban lines to a high-frequency locally owned urban
railway. It was distinctive in the design of its cars which swivel
in the middle to allow tight bends to be negotiated. In the mid-
1990s it also pioneered mobile-phone connectivity in its
tunnels.

Initially the Metro served only Tyneside, taking over some
of the conurbations' suburban railway, parts of which were
old and ageing, notably the North Tyneside Loop and the line
from Newcastle to South Shields. For example, the stretch
between South Shields and Chichester stations dated from 1834
and had been extended westwards to Gateshead in 1839. That
year also saw the opening of the line that became part of the
North Tyneside Loop (*see* 1841). The system converted from
steam to electric power in 1903, and British Rail switched to

OPPOSITE ABOVE: NEXUS, *Map of the Tyne and Wear Metro System* (2008) [NEXUS]
OPPOSITE BELOW: NEXUS, *Ruler featuring the selected Metro Car design* (1979) [AUTH]
An example of pre-opening advertising merchandise.

diesel units in the 1960s by which time the electric rolling stock had become increasingly unreliable and passenger numbers had fallen.

At that point, as plans to make Newcastle 'the Brasilia of the North' were being implemented, it was felt that there had to be significant improvements to public transport, including easing traffic congestion over the Tyne. In 1971, *The Transport Plan for the 1980s* proposed a rail solution that would provide modern, fast and reliable access to the heart of Newcastle. Voorhees & Associates were commissioned to examine alternatives for the North Tyneside Loop, comparing an LRT solution with four other options: no change, upgrade, convert to busway or close the route entirely. The viability of the LRT solution was agreed, and the Tyneside Metropolitan Railway Act was passed in 1973 and, aided by a 75 per cent grant from the central government, construction began in 1974, with the testing of prototype passenger cars on a section of new track.

The Metro opened in stages from 1980 with services running from the new underground station at Haymarket to the former British Rail station at Tynemouth via Longbenton and Whitley Bay. By the end of 1981 the southern half of the North Tyneside Loop, ending at St James beside Newcastle United's stadium after tunnelling under the city centre, was complete. So, too, were a branch from South Gosforth west to Kenton Bank Foot along the disused track of the 1906 Ponteland line, and the Haymarket to Heworth section under the city centre and across the Tyne via the Queen Elizabeth II Bridge, formally opened by the Queen in 1982. Then followed the completion of the South Shields line from Heworth in 1984, and the extension to Newcastle International Airport from Bank Foot in 1992. Finally, in 2002 the Queen was back to open the Sunderland section from Pelaw to South Hylton, making it truly the Tyne and Wear Metro. Two new stations were added at Northumberland Park and Simonside by 2008.

In its first year of full operation more than 60 million passenger journeys were made. Its popularity was only in small measure due to advance publicity, including merchandise like the 12-inch ruler that featured the selected passenger-car design. It had much more to do with it being the backbone of a fully integrated public transport network. Before the Metro, buses had not been timetabled to fit in with the suburban trains and passengers could spend a third of their journey time standing at platforms and bus stops. This all changed with the Metro, but unfortunately only temporarily because this initiative was scuppered by the deregulation of bus services in 1986.

Even so, the Metro had the supreme advantage of bringing passengers to the heart of Newcastle city centre. Previously, train stations were located on its edge, but now passengers from any direction could emerge at the foot of Grey's Monument with direct entry to the Eldon Square Shopping Centre. Interchanges like Regent Centre, Four Lane Ends and Heworth, with their multi-storey car parks, also helped to channel commuters and shoppers onto the system. This altered the pattern of shopping, not just by bolstering the city centre at the expense of district shopping centres, but also reducing the footfall along the once-vibrant streets around Grainger Street and Clayton Street.

Inevitably there are significant areas that are not served by the Metro. By largely following the existing suburban rail lines, it omits places that have seen much of the residential and employment growth since the 1970s. These include Washington New Town and Killingworth, the Quayside regeneration area along both banks of the Tyne, the Silverlink industrial zone north of the Tyne Tunnel, Doxford Park in Sunderland, and the Balliol and Gosforth estates on the way to Killingworth. Perhaps the most serious gaps, however, are the absence of lines on the western side of the conurbation, since the Metro terminates at St James, and also the lack of a link from Gateshead to the Team Valley trading estate and the MetroCentre.

The map of extensions proposed gives a clear idea of the scale of the challenge involved in plugging all these gaps. Such a big expansion of the system would require a substantial level of capital investment as well as greatly increased running costs. As it is, the system has faced increasing difficulty in maintaining a frequent and reliable service as the rolling stock and infrastructure have aged. Very welcome, therefore, was the 2019 announcement that the Department of Transport would

NEXUS, *Proposed extensions of the Metro system* (2016) [NEXUS] The 2016 Metro and Local Rail Strategy document shows proposed extensions to the Metro system in red, with plans now being brought forward for the section between Northumberland Park and Ashington.

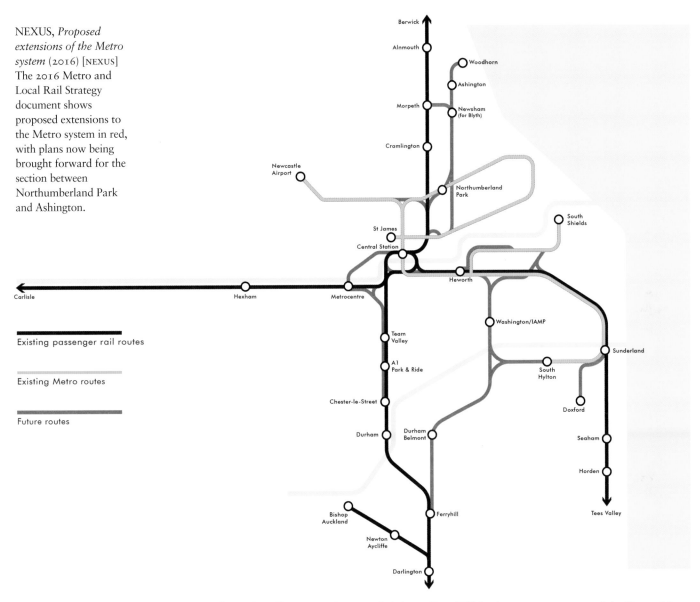

Berwick

Alnmouth

Woodhorn

Ashington

Morpeth

Newsham (for Blyth)

Cramlington

Newcastle Airport

Northumberland Park

St James

South Shields

Central Station

Heworth

Carlisle

Hexham

Metrocentre

Washington/IAMP

Team Valley

Sunderland

A1 Park & Ride

South Hylton

Chester-le-Street

Doxford

Durham

Durham Belmont

Seaham

Horden

Bishop Auckland

Ferryhill

Tees Valley

Newton Aycliffe

Darlington

Existing passenger rail routes

Existing Metro routes

Future routes

provide £337 million towards a new fleet of 42 five-car articulated trains and a new depot at Howdon, due to come into service in 2024. The Government's March 2020 Budget also set aside almost £100 million for track investment between Pelaw and Bede stations, allowing an extra 30,000 passenger journeys each day.

One incidental benefit of Metro construction was that much of the material removed from under the city centre was used to heighten Cow Hill in the western corner of the Town Moor (itself an artificial creation initially from the mining spoil generated by the local pits and then from material excavated in the construction of the 'north-west radial' road that passes nearby towards Cowgate). Now the hill not only permits a panoramic view of the city but also provides a popular slope for tobogganing, though increasingly rarely in these days of global warming.

2020

The ups and downs of Newcastle United

What is a club in any case? It's the noise, the passion, the feeling of belonging, the pride in your city . . . It's a small boy clambering up the stadium steps for the very first time . . . and, without being able to do a thing about it, falling in love.

Sir Bobby Robson,
Newcastle Evening Chronicle, 2013

As the bird's-eye view over this part of the city dramatically illustrates, St James' Park, the home of Newcastle United Football Club (NUFC), looms not only over its immediate surroundings (not least the splendid Leazes Terrace designed by John Dobson in the 1820s), but is also a visible landmark from much of the city's hinterland. This physical domination is matched by the club's continuous presence in the local media and the emotional impact of its fortunes among its faithful supporters. It has often been said that a weekend win led to higher productivity in the following week. Despite the team's relative lack of success over the last half-century, the ground and the football team remain an integral part of Newcastle's image, instantly recognisable and provoking a plethora of associated feelings and memories.

As the name suggests, Newcastle United was formed from the union of two former clubs, Newcastle East End and Newcastle West End, although the first club to play at St James' Park was called Newcastle Rangers, formed in 1878. The latter club was initially forced to play in Gateshead as it was unable to find a ground in Newcastle, but it moved to a piece of

---

OPPOSITE: Newcastle University Estates Office, *Bird's-Eye View of St James' Park, City Centre and Bridges* (2020) [NEWCASTLE UNIVERSITY ESTATES OFFICE] St James' Park is in the left foreground; the Helix development to its right, although this is an artist's impression superimposed onto the photograph; the colourful circular building to the right in the middle distance is the Centre for Life; six of the seven bridges are visible at the top.

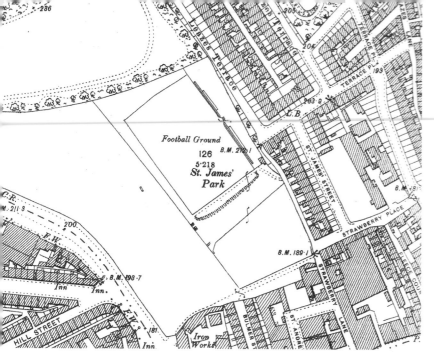

Ordnance Survey, *St James' Park* (1898) [AUTH] 25-inch map, second edition, 1898, Tyneside Sheet II.

End ceased to operate as a football club. Their players and officials were given the chance to join the East End club. On Friday 9 December 1892, East End became Newcastle United, playing their first game at St James' Park (impressively, against Glasgow Celtic) in front of 6,000 spectators. The 1898 map shows the sparse amenities of that time, and even by 1914 facilities were still minimal despite the fact that the club had joined the Football League (Second Division) in 1893 and enjoyed a remarkable period of success. Between 1904 and 1910 the club won three League titles and the FA Cup. It was to win a fourth League title in 1926–27 and the FA cup no less than a further five times. However, the last occasion was in 1955, although one of its greatest achievements was in 1969 when it won Europe's Inter-Cities Fairs Cup.

As football has become more significant in commercial and image terms, St James' Park has had an ever more turbulent history. At the start of the twentieth century, the ground's capacity was 30,000, but redevelopment by 1905 increased this substantially. At that time, only the west stand was covered, and for the next 60 years the only significant change was the construction of a metal roof over the Leazes end in 1927. Despite an ambitious plan in the 1920s, the other terraces remained open to the weather until the 1970s. The only significant change in the early post-war period was the installation of floodlights in 1959, proudly displayed on the 1961 programme cover, and still in 1970 there were only 4,700 seats in a stadium with a capacity of well in excess of 50,000.

The cheery magpie mascot with top hat and cigar is a reminder of an apparently less sophisticated era

Newcastle United, *Newcastle United vs Stockport County* (1961) [AUTH] Programme for the FA Cup game on 1 February 1961. Newcastle won 4–0.

enclosed ground in front of Leazes Terrace in 1880, which immediately took the name St James' Park. The lease of this ground was then taken over by the West End club which had been formed in 1882 (ironically, as an offshoot of a cricket club). This club had a somewhat nomadic existence, playing initially on the edge of the Town Moor just off Claremont Road and then, when ball games were temporarily banned from the Town Moor in 1885, moving to a site in Jesmond adjacent to the Great North Road. It moved to St James' Park in May 1886 and erected an eight-foot-high fence around the field.

Meanwhile, Newcastle East End had itself been formed from the amalgamation in 1881 of two clubs, the very short-lived Rosewood and the Stanley who played on vacant land near Stanley Street in Byker. Remarkably, this latter club had also been an offshoot of a pre-existing cricket club. The new club moved from Stanley Street to Chillingham Road in Heaton in 1884 and proved to be far more successful than their West End rivals at St James' Park who, by the early 1890s, were almost insolvent. In fact, so perilous was the latter's position that they approached the East End club in 1892, requesting that it should take over the lease of St James' Park for West End's remaining 12 years and offering a donation of £100 to the East End finances to do this. This was agreed and West

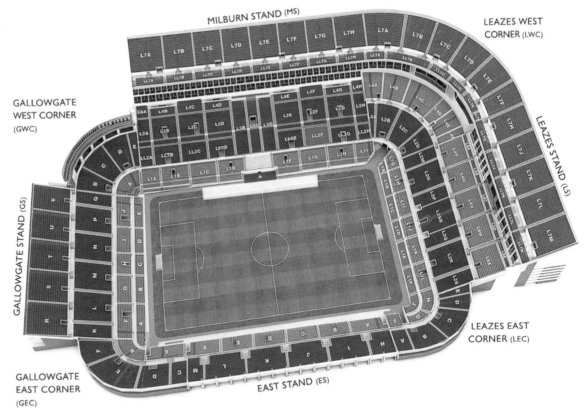

Newcastle United, *Seating Plan, Newcastle United Stadium* (2020) [NUFC] Newcastle Stadium as it is today.

MILBURN STAND (MS)

LEAZES WEST CORNER (LWC)

GALLOWGATE WEST CORNER (GWC)

LEAZES STAND (LS)

GALLOWGATE STAND (GS)

LEAZES EAST CORNER (LEC)

GALLOWGATE EAST CORNER (GEC)

EAST STAND (ES)

but one with considerably less angst. By contrast, the 1960s saw the start of a much more unsettled period. By the early 1960s the current lease from the City Council was running out (the main reason why St James' was not selected as a venue for the 1966 World Cup games). The Council, led by the charismatic T. Dan Smith, saw the site as an ideal location for a large-scale, community-based sports complex with swimming pools, badminton courts, five-a-side pitches and much more. This potential threat caused Newcastle United's Board to look elsewhere for a ground, and they settled on a 35-acre site off Sandy Lane at Gosforth, near the racecourse. Ambitious plans were drawn up and, in 1968, they applied for planning consent. However, the club became increasingly concerned about the cost of the move, and the City Council had become less confident of its large-scale planning approach. As a result, both sides were eventually happy to sign a new 99-year lease for St James'.

By 1987 a long-awaited redevelopment scheme began, in part enforced by public safety regulations introduced after the Bradford City fire disaster. The old West Stand was at last replaced by the Milburn Stand and the Sir John Hall Stand replaced the old Leazes End terraces in 1993. The rest of the ground was renovated, producing a 37,000 all-seater stadium. Between 1998 and 2000 soaring double tiers were added to both these stands, shown to good effect in the seating plan, which brought the ground capacity up to the present 52,388.

The latter expansion was actually a compromise solution after the club, led by the entrepreneur Sir John Hall, proposed to move the ground to a new location, taking over a significant part of Leazes Park with an associated multi-sports complex and commercial development. It was intended that the capacity of the stadium would eventually be increased to 90,000. Reflecting the intensity of feeling towards the existing site, a public protest with a 38,000-signature petition led to the demise of the scheme. In a fit of pique, Hall threatened to move the ground to Gateshead, thereby surrendering most of the goodwill he had enjoyed with fans. The purchase of Newcastle United by another entrepreneur, Mike Ashley, in 2007 has led to even more instability, with Ashley attempting to sell the club on at least four occasions.

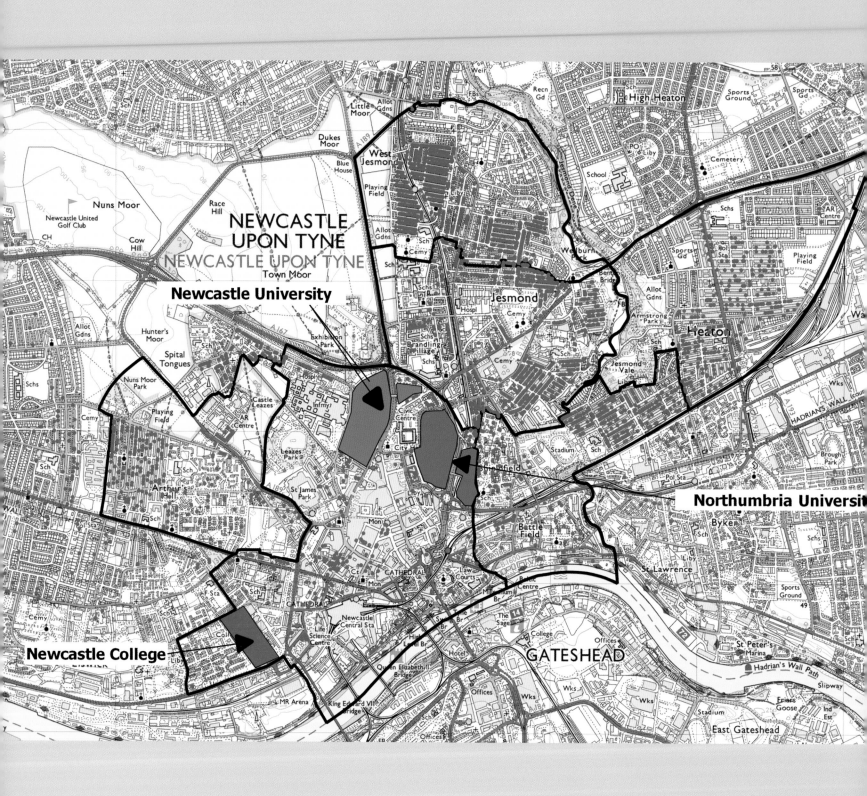

NEWCASTLE
UPON TYNE
NEWCASTLE UPON TYNE
Town Moor

Newcastle University

Northumbria Universit

Newcastle College

GATESHEAD

2021a

University city and the civic university

The pre-war plans for an educational complex around Barras Bridge and the Haymarket have been translated beyond the modest expectations of the 1930s . . . while the economy has come to depend upon the numbers of students, lecturers and researchers . . . they have played a central role in Newcastle's social and cultural renaissance.

A.W. Purdue, *Newcastle: The Biography*, 2011

Although Tyneside has long been a centre of learning and training, it was only in 1963 that Newcastle became a university city in its own right. This represented the culmination of a continuous record of post-school education spanning nearly two centuries and building on an even longer history of scientific inquiry and innovation. With the waning of the manufacturing sector, higher education (HE) has become one of the main pillars of the local economy, and its strong recent growth is making it an increasingly visible component of the city landscape in both physical and human terms. Its universities are now as much 'of the city' as 'in the city'.

The City Council's Development Plan of 1945 proposed that the northern fringe of the city centre should be developed as a university precinct focused on what had become King's College in 1937. The city's 1963 plan reinforced this educational strategy, with a sizeable expansion of the site of what that year had become the University of Newcastle upon Tyne. As well as extending south to St Thomas Street, it now crossed Claremont Road and linked up with the 'further education precinct', extending down the eastern side of the Civic Centre south to Ellison Place. Here Rutherford College (established in 1880), was soon to merge with two other colleges to form Newcastle Polytechnic, later Northumbria University.

OPPOSITE: Colin White, *Student properties in six 'High Student Wards'* (2020/21) [NCC] These properties, exempt from council tax, are close to the two university campuses principally in Jesmond, Heaton and Sandyford.

Simon Earp, *Purpose-Built Student Accommodation Blocks* (2021) [NCC] Already a large number completed (in blue) with more under construction (yellow) or with planning permission (pink).

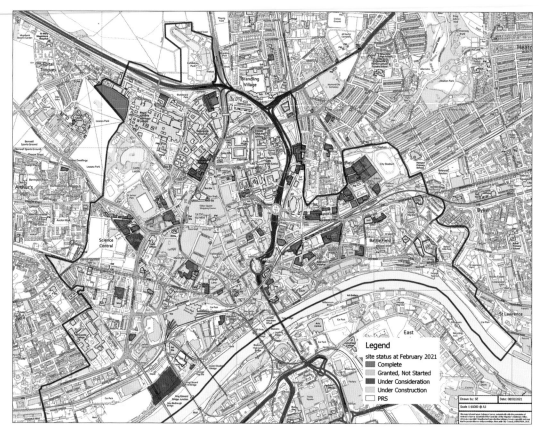

Higher education in Newcastle has vastly expanded over the past half century. Newcastle University's 1963 plan envisaged student numbers rising from around 4,000 to some 6,000 with the possibility that 'it may even expand beyond this'. By the academic year 2010/11 its total enrolment had increased to almost 21,000, while Northumbria's was over 29,000, and a decade later the total number of HE students in the city reached the 60,000 mark, including some 3,000 registered for HE courses at Newcastle College, such that they now comprise one in five of the city's population. This phenomenal rise in student numbers, paralleled by massive growth in the universities' research activity, has had a major impact on the city in terms of the provision of student housing and the development of new university buildings.

The two maps of the provision of residential accommodation for these students dramatically indicate how the city has been altered for this purpose. The red dots on the main map indicate buildings which were built as family homes with at least two bedrooms but are now occupied entirely by students and therefore identifiable by virtue of not being eligible for council tax. One in every fifteen homes across the city is now a student property, the highest proportion in the country along with Exeter. Clearly, these are not evenly distributed across the city, being concentrated in inner neighbourhoods within a mile of the two main university campuses.

Student penetration has been greatest in Jesmond, where the predominantly middle-class population has been subjected to large-scale 'studentification' and Jesmond's central spine of Osborne Road has emerged as a nightclubbing 'hotspot' rivalling the famous Bigg Market in the centre of town and a gentrifying Quayside, serving to reduce the area's attractiveness for families. Similar transformation has taken place further east in Heaton and also westwards in Arthur's Hill and Spital Tongues, the latter close to the hall of residence at Castle Leazes, the only survivor of Newcastle University's three main halls that – along with its student flats on Richardson Road – accommodated a significant proportion of its student growth until well into the 1980s.

Since the 1990s, however, the main change in student housing provision has taken the form of blocks of purpose-built student accommodation (PBSA), which has reinforced the concentration of student quarters in the heart of the city. A City Council assessment of 2007 identified a total of 47 sites potentially suitable for these apartment blocks, two-thirds of which were located within or on the immediate fringes of the city centre. By 2021 most of these had come to fruition, as indicated in blue on the map, with a couple of further schemes under construction (yellow) and some others for which planning permission has been granted but building has not yet started (pink). The map also shows two large new sites which in February 2021 were being considered for PBSA schemes (purple), one adding to the recently completed blocks adjacent to St James' Park football stadium and an even larger site on the Quayside.

These PBSA developments are all in addition to the steady expansion of the university campuses themselves, with the 'university precinct' of the 1945 Plan extending southwards on both sides of the city's main commercial core. It now crosses the Central Motorway to embrace not just Northumbria's new East Campus but also Shieldfield, King's Manor and as far as City Road just above the Quayside. On the west side of the city centre, it has leapt over St James' Park football stadium to embrace the cleared site of the former Tyne (and then Scottish and Newcastle) brewery. Yet even this wider zone fails to encompass the full scale of higher education and research activity that includes Newcastle College (which started awarding degrees in the 2000s), the International Centre for Life near Central Station and the Campus for Ageing and Vitality on the site of the former General Hospital, let alone facilities further afield like Northumbria's Coach Lane Campus to the north-east and Newcastle University's Dove Marine Laboratory at Cullercoats and Cockle Park Farm in Northumberland.

Currently, it is the Helix – being developed on the former brewery site – that is the most eye-catching new scheme. Resulting from Newcastle's designation in 2005 as a Science City, a clutch of imaginatively designed buildings has been soaring skywards over the last few years. The first buildings comprised 'the Core', accommodating a mix of occupants including cloud computing, professional development and firm incubator space; the Urban Sciences Centre including the Department of Computing Science and Institute for Sustainability; and the National Innovation Centre with three national centres specialising in data science, ageing and health. This huge scheme has involved a partnership between Newcastle University, the City Council, the regional development agency One North East (until its abolition in 2011) and the insurance company Legal & General. The Helix aims not only to attract scientific endeavour but also to create the 'quadruple helix' of partnership between academia, the public sector, private business and the wider community, as promoted by the civic university movement.

The result of all these changes is a virtual 'colonisation' of the central and inner parts of the city. While this scale of development has generated challenges, it has begun to reduce some pressure on the traditional student areas. Town–gown relations have also improved as a result of the two universities adopting such initiatives as the funding of extra police presence in these areas, known as Operation Oak, and the annual 'best student neighbour' awards, though this progress has been negatively affected by reports of students ignoring Covid-19 lockdown rules. The pandemic has also raised questions about the future need of both student accommodation and university buildings, if the enforced switch to student online learning and staff home working leads to HE changing its mode of operation towards something more akin to the Open University. There is also uncertainty about the long-term effect of Brexit on the recruitment of international students, especially from the remaining 27 countries of the European Union.

2021b

Iconic Tyne bridges

Few people fail to be impressed by the bridgescape over the tidal Tyne between Gateshead and Newcastle, essentially because of the variations in the forms of bridge built along a few hundred yards of river . . . This variation in bridge form stems from differing functions, materials, building techniques and site constraints.

S. Linsley, in Archer, *Tyne and Tide*, 2003

We began this book with the origins of Newcastle as the Romans' lowest bridging point of the Tyne nearest the sea. To complete the circle of history, we end it with an account of Newcastle's iconic bridges. If Rome is a city of seven hills, Newcastle is a city of seven bridges. We use Open Street Map (OSM), cartography based on digital technology, to introduce these crossings. The impressive detail of the OSM map captures the equally impressive collection of bridges between Newcastle and Gateshead and gives a sense of the complexity involved in these various manifestations of Tyne crossings.

The Tyne gorge presented both challenges and opportunities to bridging the river: challenges, because the steepness of the banks made low-level bridges less than ideal, while high-level bridges needed very deep piles driven into the shifting sands at the bottom of the river before hitting solid foundations; and yet opportunities since, if the difficulties could be overcome, the sheer height of the bridges held out the promise of impressive structures. The most dramatic of the bridges are those linking the hearts of the two towns: the High Level, the Swing, the Tyne and the Gateshead Millennium.

For long, there was only a single bridge linking Gateshead and Newcastle: initially a wooden Roman bridge built on stone

OPPOSITE: Open Street Map, *Newcastle–Gateshead Seven Bridges* (2021) [OSM] In 1864 a proposal from the Tyne Improvement Commission to dig a tunnel between Newcastle and Gateshead, from where the current Tyne Bridge starts to the Gateshead end of the High Level Bridge, was given serious consideration.

OLD TYNE BRIDGE AS IT STOOD IN 1739.

TYNE BRIDGE TAKEN DOWN 1866-67.

NEW HYDRAULIC SWING BRIDGE OPENED 1876.

HIGH LEVEL BRIDGE

REDHEUGH BRIDGE

ABOVE: M. Lambert, *Five Historic Tyne Bridges* (1874) [SANT] From Lambert's *Newcastle Memorandum*, an 1874 almanac.

OPPOSITE: Newcastle Corporation, *High Level Bridge and Proposed Swing Bridge* (1864) [LIT & PHIL] In 1924 the Swing Bridge opened for around 6,000 vessels; in 2000 it had to open for only nine.

piers and named the *Pons Aelius* after the family of Emperor Hadrian. It crossed the river on a line almost exactly where the Swing Bridge now stands. In the eleventh century the Normans built a new wooden superstructure on the Roman piers. This was destroyed in a massive fire in 1248 but was replaced by a new stone bridge (again incorporating the Roman piers) with gates at both ends, a tower in the middle, a chapel at its northern end and a variety of houses along both sides. This, in turn, was irreparably damaged by the Great Flood of 1771 and replaced by a new stone bridge, built between 1773 and 1781. This stood as the lone low-level crossing between the two towns until it was replaced by the Swing Bridge.

The High Level Bridge was opened in 1849 by Queen Victoria and celebrated by masses of excited spectators. It was a bold construction designed by Robert Stephenson. There had been several similar proposals, most notably by the architects John and Benjamin Green a decade earlier, but Stephenson's bridge had been fought for by the swashbuckling 'Railway King', George Hudson, as it was essential for the completion of the rail connection to Edinburgh. Stephenson's High Level Bridge became the most dramatic of all the Tyne bridges thus far, constructed with twin decks, catering for trains above and road users below and with enclosed pedestrian pavements at the sides of the road. The upper and lower levels are joined by elegant cast-iron arches, and the whole structure stands proudly on massive high pillars which allow the passage of sea-going ships. Tolls were charged on the lower deck for both traffic and pedestrians. From the 1880s horse-drawn bus services operated across the bridge, and from the 1920s tramcars began to use it. Tolls eventually stopped being charged in 1937.

The Swing Bridge replaced the old arched stone bridge which had become an increasingly unsustainable obstacle to river traffic. The Tyne Improvement Commission's engineer, J.F. Ure, was centrally involved in this replacement, having as early as 1860 suggested 'a plan that would enable the two centre spans to be opened for the passage of masted ships', and he appears to have later collaborated in the design of the Swing Bridge with Sir William Armstrong. Work began in 1873, and it opened for traffic in 1876. Armstrong clearly had a vested

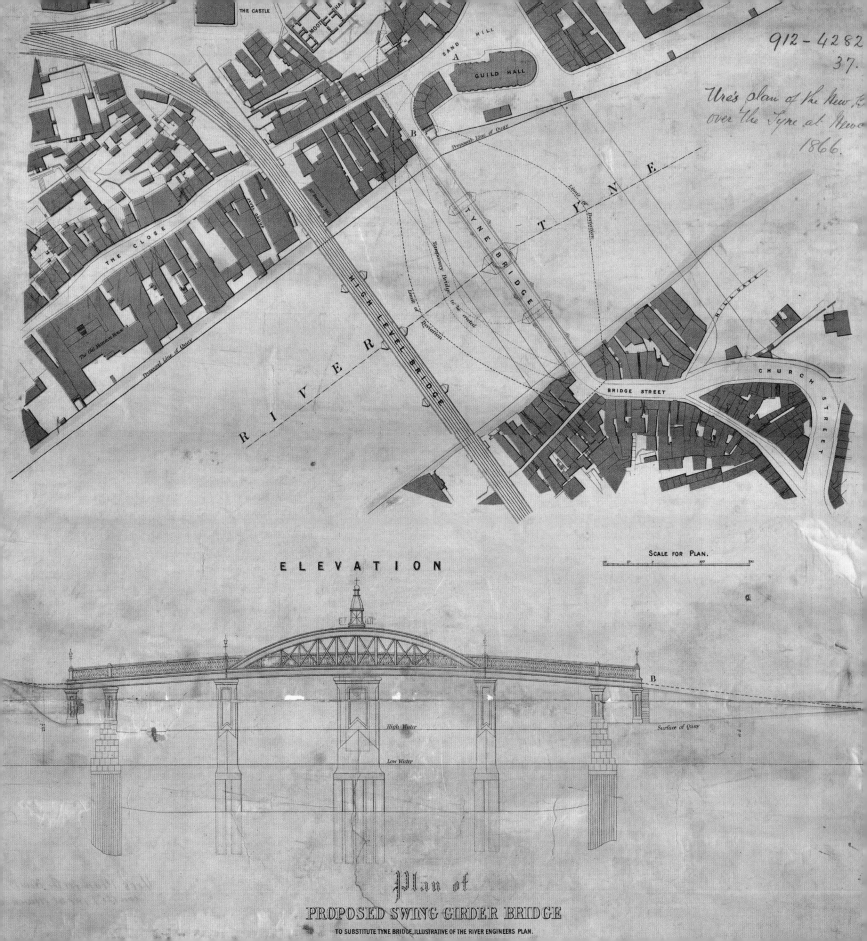

912-4282
37.

Ure's Plan of the New B...
over the Tyne at New...
1866.

THE CASTLE

SAND HILL

A

GUILD HALL

B

Proposed Line of Quay

T Y N E

TYNE BRIDGE

Limit of Deviation

THE CLOSE

HIGH LEVEL BRIDGE

Temporary Bridge to be erected

Limit of Deviation

R I V E R

MILL GATE

BRIDGE STREET

CHURCH STREET

SCALE FOR PLAN.

E L E V A T I O N

B

High Water

Surface of Quay

Low Water

Plan of
PROPOSED SWING GIRDER BRIDGE
TO SUBSTITUTE TYNE BRIDGE, ILLUSTRATIVE OF THE RIVER ENGINEERS PLAN.

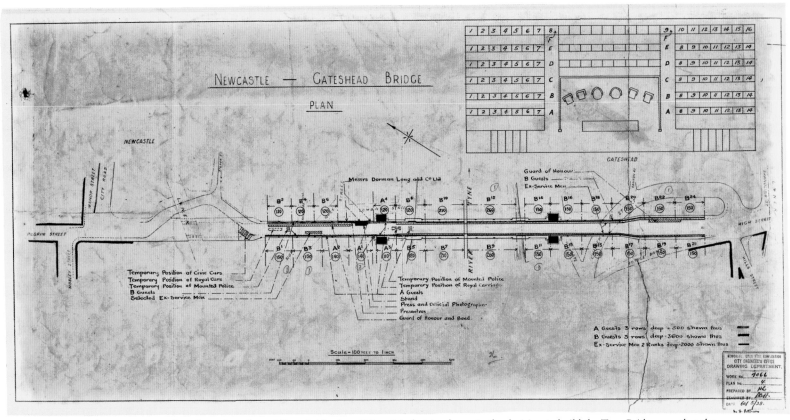

Newcastle City Engineer, *Royal opening of 1928 Tyne Bridge* (1928) [GLH] A decisive factor in the decision to build the Tyne Bridge was that the Ministry of Transport met 60 per cent of the cost.

interest in ensuring that river traffic could readily reach his massive Elswick engineering works. Indeed, the first ship (rather ironically called the *Europa*) to pass under the bridge collected a huge gun from Elswick that had been commissioned by the Italian government. The ever-inventive Armstrong had especial expertise in hydraulic systems and he used hydraulic engines in the central pier to swing the bridge parallel to the river, thereby creating two wide channels through which river traffic could pass. It is a charming and very distinctive structure. The control room sits in a quaint turret with a walkway around its windows and surmounted by a large cupula housing a lantern.

The Tyne Bridge was built in response to the growth in road traffic. It was formally opened in 1928 by King George

V and originally carried the Great North Road up to Edinburgh. The plan above evokes the careful and detailed preparations for this royal event. Some 28,000 schoolchildren were given a holiday but they must have lined the route as, on the bridge itself, adult guests were meticulously placed. Two thousand ex-serviceman lined the bridge two ranks deep, 500 category 'A' guests were located around the royal stand at the Newcastle end (shown in the inset, top right) and 3,600 category 'B' guests were strung along both sides of the bridge, three rows deep. For ease of access and dispersal, the guests were further divided into smaller groups ranging from 110 to 260. The King's speech was the first to be recorded by Movietone News and the ceremony was ended with a 21-gun salute and a peal of bells.

The Tyne Bridge is a single-span steel structure with a single parabolic arch. At each end there are huge granite towers, designed by local architect Robert Burns Dick. They were originally intended as warehouses built over five storeys, but instead they housed lifts. At one time the lift at the Newcastle end could be used to travel between the Quayside and the upper levels of the town. Burns Dick's original design was even more monumental, modelled on Egyptian temple entrances and would have provided an impressive ceremonial entry to the town. The similarity to the Sydney Harbour Bridge is obvious, with both designed by the engineering firm of Mott, Hay and Anderson. As the Tyne Bridge was completed four years before Sydney, many Tynesiders argue that it was the model for the Australian 'version'. However, the Sydney bridge was started more than a year before the Tyne so it was clearly not a replica. At one time the bridge carried trams as well as motor traffic, although it was never used by trolley buses since Gateshead did not develop a trolley-bus network. The timing of the bridge's building was significant, with its construction offering some help in relieving the high unemployment of the 1920s. The Tyne Bridge is probably the most celebrated of the bridges, used as an instantly recognisable icon of Newcastle and Tyneside.

Opened in 2001, the Gateshead Millennium Bridge is the latest and the first to sit downstream of the Tyne Bridge. It can be thought of either as Newcastle's seventh bridge or, more realistically, as Gateshead's first, built as part of the striking regeneration of Gateshead's river frontage. It is a pedestrian and cycle bridge with a mechanism operated by six hydraulic rams that rotate the bridge back to allow river traffic to pass beneath. To open and close it takes just four minutes. It is the world's first tilting bridge and has won numerous awards. Known popularly as 'the blinking eye', it rapidly gained the affection of locals and visitors, and its arresting design has added an extra aura to views of the river. It has also allowed pedestrians crossing over it to complete a low-level circuit around both sides of the river, using the Swing Bridge.

Newcastle's three other bridges are somewhat less dramatic, but each contributes to the splendid panorama of river cross-

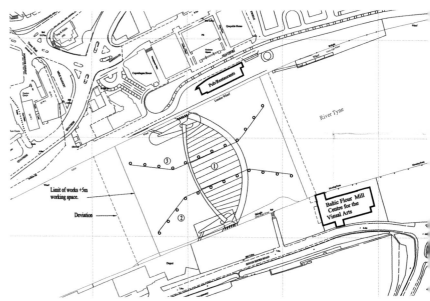

M. Godfrey, *Gateshead Millennium Bridge Works Plan* (1998) [GLH] Even allowing for inflation, at £22 million the Gateshead Millennium bridge is the most costly to be built on the Tyne.

ings. Over time the Redheugh Bridge has provided three successive road crossings, the first opened in 1871, the second in 1901 and the third in 1983. It began as a toll bridge with revenues benefitting from traffic congestion on the narrow roadway of the High Level. With the opening of the toll-free Tyne Bridge, however, toll revenue suffered, and it was bought by Newcastle and Gateshead in 1937 and the tolls abolished. Weight limits and access restrictions led to its replacement by the current Redheugh Bridge that opened in 1983.

The King Edward VII Bridge is a railway bridge opened in 1906 with four railway tracks that eased congestion on the approach to Newcastle (although as early as 1841 a similar bridge had been mooted). It is the main line from which the dramatic views of the High Level, Swing and Tyne bridges are seen by travellers to and from London. Finally, the utilitarian Queen Elizabeth II Bridge, opened in 1981, was built specifically to carry Tyne and Wear Metro trains across the river. The Metro line travels in tunnels as it approaches both ends of the bridge but emerges into the open to cross the bridge.

Further reading

CARTOGRAPHY

Arribas, A., 'The Conservation Treatment of a Plan of Newcastle upon Tyne by James Corbridge', *Archaeologia Aeliana*, Series 5, 29 (2001) pp.293–7.

Barke, M. and Buswell, R.J. (eds), *Historical Atlas of Newcastle upon Tyne* (Newcastle upon Tyne: Newcastle Polytechnic, 1980).

Barke, M. and Buswell, R.J. (eds), *Newcastle's Changing Map* (Newcastle upon Tyne: Newcastle upon Tyne City Libraries & Arts, 1992).

Clavering, E. and Rounding, A., 'A map and its meaning', *Archaeologia Aeliana*, Series 5, 40, (2011), pp.243–58.

Delano-Smith, C. and Kain, J.P., *English Maps. A History* (London: The British Library, 1999).

Frostick, R., 'James Corbridge and his Plan of Newcastle upon Tyne, 1723', *Archaeologia Aeliana*, Series 5, 32 (2003), pp.171–8.

Graham, F., *Maps of Newcastle* (Newcastle upon Tyne: Frank Graham, 1984).

Honeyman, H.L. and Jones, M.E., 'Thomas Oliver and his Plans for Central Newcastle', *Archaeologia Aeliana*, Series 4, 29 (1951), pp.239–52.

Robson, B.T., 'Friends in the north: Infighting in the re-making of Newcastle', *Northern History*, 53, 1 (2016), pp.78–97.

Scott, M., *Rook's 1827 Plan of North Shields and Tynemouth* (North Shields: Old Low Light, 2017).

Spence, C.J., 'Notes on the Plates and Maps of the Tyne in Gardner's *England's Grievance Discovered* of 1655', *Archaeologia Aeliana*, Series 2, 8 (1889), pp.285–305.

Wake, T., 'Isaac Thompson's Plan of Newcastle upon Tyne, 1746', *Archaeologia Aeliana*, Series 4, 14 (1937), pp.110–22.

Welford. R., 'The Walls of Newcastle in 1638', *Archaeologia Aeliana*, Series 2, 12 (1887), pp.230–5.

Whitaker, H., *A Descriptive List of the Maps of Northumberland, 1576–1900* (Newcastle upon Tyne: Society of Antiquaries of Newcastle upon Tyne, 1949).

GENERAL

Charleton, R.J., *The History of Newcastle upon Tyne* (Newcastle upon Tyne, 1885).

Colls, R. and Lancaster, B. (eds), *Newcastle upon Tyne: A Modern History* (Chichester: Phillimore, 2001).

Fraser, C.M. and Emsley, K., *Tyneside* (Newton Abbot: David & Charles, 1973).

Jackson, D., *The Northumbrians. North-East England and its People. A New History* (London: Hurst & Company, 2019).

Mackenzie, E., *A Descriptive and Historical Account of the Town and County of Newcastle upon Tyne* (Newcastle upon Tyne, 1827).

Manders, F.W.D., *A History of Gateshead* (Gateshead: Gateshead Corporation, 1973).

McCord, N., *North East England. The Region's Development* (London: Batsford, 1979).

Middlebrook, S., *Newcastle upon Tyne: Its Growth and Achievement* (Newcastle upon Tyne: Newcastle Chronicle, 1950).

Newton, D. & Pollard, A.J. (eds), *Newcastle and Gateshead Before 1700* (Chichester: Phillimore, 2009).

Purdue, A.W., *Newcastle: The Biography* (Stroud: Amberley Publishing, 2011).

Robinson, F. (ed.), *Post-Industrial Tyneside* (Newcastle upon Tyne: Newcastle upon Tyne City Libraries & Arts, 1988).

THEMATIC

Archer, D., *Tyne and Tide: A Celebration of the River Tyne* (Ovingham: Daryan Press, 2003).

Armstrong, C., *Tyneside in the Second World War* (Chichester: Phillimore, 2007).

Barke, M., 'The middle-class journey to work in Newcastle upon Tyne, 1850–1913', *Journal of Transport History*, Third Series, vol. 12 (1991), 2, pp.107–34.

Barke, M., ' "The Devouring Element": The Fire Hazard in Newcastle upon Tyne, 1720–1870', *The Local Historian*, vol. 43 (2013), 1, pp.2–13.

Barke, M., 'The North East Coast Exhibition of 1929: Entrenchment or Modernity?', *Northern History*, 50 (2014), 1, pp.149–72.

Barke, M., 'The development of public transport in Newcastle upon Tyne and Tyneside, 1850–1914', *Journal of Local and Regional Studies*, vol. 12 (1992), 1, pp.29–52.

Baron, F., *The Town Moor Hoppings* (Newbury: Lovell Baines Print, 1984).

Benwell Community Project, *Private Housing and the Working Class*, Final Report Series, No. 3, 1978.

Benwell Community Project, *The Making of a Ruling Class*, Final Report Series, No. 6, 1978.

Burns, W., *Newcastle. A Study in Replanning at Newcastle upon Tyne* (London: Leonard Hill, 1967).

Clarke, J.F., *Building Ships on the North East Coast: a Labour of Love, Risk and Pain*. 2 vols (Newcastle upon Tyne: Bewick Press, 1997).

Conzen, M.R.G., 'The plan analysis of an English city centre', pp.25–53, in Whitehand, J.W.R. (ed.), *The Urban Landscape: Historical Development and Management* (London: Academic Press, 1981).

Dendy, F.W., 'The six Newcastle chares destroyed by the fire of 1854', *Archaeologia Aeliana*, Series 2, 18 (1895–96), pp.241–57.

Faulkner, T., 'The early nineteenth century planning of Newcastle upon Tyne', *Planning Perspectives*, 5 (1990), pp.149–67.

Faulkner, T., 'Conservation and renewal in Newcastle', in Faulkner, T. (ed.), *Northumbrian Panorama* (London: Octavian Press, 1996).

Faulkner, T. and Greg, A., *John Dobson. Architect of the North East* (Newcastle upon Tyne: Tyne Bridge Publishing, 2001).

Faulkner, T., Beacock, P. and Jones, P., *Newcastle & Gateshead. Architecture and Heritage* (Liverpool: The Bluecoat Press, 2006).

Graves, C.P. and Heslop, D.H., *Newcastle upon Tyne: The Eye of the North. An Archaeological Assessment* (Oxford: Oxbow Books, 2013).

Guthrie, J., *The River Tyne: Its History and Resources* (London: Longmans, 1880).

Hoole, K., *Tomlinson's North Eastern Railway: Its Rise and Development* (Newton Abbot: David & Charles, 1967).

Hoole, K., 'Railway electrification on Tyneside, 1902–67', *Transport History*, 2 (1969), pp.258–83.

Howell, R., *Newcastle upon Tyne and the Puritan Revolution* (Oxford: Clarendon Press, 1967).

Johnson, G.A.L. (ed.), *Robson's Geology of North East England: Transactions of the Natural History Society of Northumbria*, 56, 2, (1995), Newcastle upon Tyne.

Lancaster, W. (ed.), *A History of Working Class Housing on Tyneside* (Newcastle upon Tyne: Bewick Press, 1994).

Mackenzie, P., *W.G. Armstrong: A Biography* (Newcastle upon Tyne: Longhirst Press, 1983).

McCord, N., 'The making of modern Newcastle', *Archaeologia Aeliana*, Series 5, 9 (1981), pp.333–46.

Pendlebury, J., 'Alas Smith and Burns? Conservation in Newcastle upon Tyne City Centre 1959–1968', *Planning Perspectives*, 16 (2001), pp.115–41.

Perry, M., *The Jarrow Crusade: Protest and Legend* (Sunderland: University of Sunderland Press, 2005).

Rennison, R.W., *Water to Tyneside. A History of the Newcastle & Gateshead Water Company* (Newcastle upon Tyne: Newcastle and Gateshead Water Company, 1979).

Serdiville, R. and Sadler, J., *The Great Siege of Newcastle 1644* (Stroud: The History Press, 2011).

Turley, R., 'Early Victorian city planning: the work of John Dobson and Richard Grainger at Newcastle upon Tyne', *Architectural Review*, 99 (1946), p.145.

Turnbull, L., *Railways before George Stephenson* (Oxford: Chapman Research Publishing, 2012).

Wilkes, L. & Dodds, G., *Tyneside Classical. The Newcastle of Grainger, Dobson & Clayton* (London: John Murray, 1964).

Wright, P.D., *Life on the Tyne. Water Trades on the Lower Tyne in the Seventeenth and Eighteenth Centuries, a Reappraisal* (Farnham: Ashgate, 2014).

Index

Brandling, Robert (1498–1568) landowner, mayor 102

Brandling, Robert William (1775–1849) landowner, railway proprietor 110–11

breweries 39, 118, 241

Bristol (1673 plan) 47

Britannia Magna (1661) 32

Britannia Romana (1732) 2

British Association (1863) 125

British Electric Traction Ltd 177

Brockett, John Trotter (1788–1842) antiquarian 66, 157

Brockley Whins, junction 110

Brooks, William A. (1802–1877) civil engineer 124, 133

Brown, Paul J., commercial artist 150

Bruce, John Collingwood (1805–1892) antiquarian 2–3

Brunel, Isambard Kingdom (1806–1859) engineer 111

Bryson, Thomas (1805–1867) town surveyor 120

Buck, Samuel (1696–1779) artist and engraver 52–5, 93

Buddle, John (1773–1843) mining engineer 173

Bull Park, Town Moor 115, 138

Bull, William (1863–1931) London planner 181

Bulman, Job James (1744–1818) banker 203

BUPA Great North Run (1999), cartoon 222

burgage plots 33, 71, 90

Burns Dick, Robert (1868–1954), architect, City Centre Plan (1924) 154–6, 179–81; City Green Belt (1927) 181–2; Tyne Bridge 247

Burns Dick Plan for Central Newcastle (1924) 155, 178, 180

Burns, Wilfred (1923–1984) town planner 207–9

Butcher Market 94, 96

Byker 28, 100, 176, 236; redevelopment 209

Cackett, Burns Dick & Mackellar, architects 154, 179

Cackett, James Thoburn (1860–1928), architect 154, 180

Camden, William (1551–1623) antiquarian x

canals, schemes 41, 80–3

Capone, William Henry, engraver 7

Carliol Croft 90

Carliol Square 106, 111

Carlisle 82, 89; railway 83, 106, 109

Carlisle, 5th Earl, Longbenton estate 59

car parks 213, 232

Casson and Berry (Manchester) 47

Casson, William, surveyor 28

Castle Garth 17, 57, 94

Castle Keep: Sections and Elevations (c.1847) 16

Castle Leazes 94, 114, 137; student residences 241

castle, Newcastle 2, 3, 13–17, 56, 153

Central Exchange 96, 153

Central Motorway 114, 208, 224, 241

Central Station 94, 97, 110, 180, 213, 241

Chapman, Clarke, engineers 169

Chapman, William (1749–1832) canal engineer 63, 80–3

chares 64–7

Chares destroyed by the Great Fire, 1854 (1855) 67

Charles I 35–7

Charleton, R.J., historian 66, 113, 115

Charlotte Square 71

Charlotte Street 79

Chartist, demonstration 115

chemical works 75, 85, 87, 128

Chinatown (Stowell Street) 220

cholera epidemics 118–19

Chorographia (1649) 35, 60

Church of England Commissioners 212

City Beautiful movement 155

City Green Belt: Plan illustrating the proposals of the Newcastle Society (1927) 181

City Road 87, 97, 241

Civic Amenities Act (1967) 208

Civic Centre 153–5, 180

Civil War 14, 34–7, 41

Clapham, Anthony, soap maker 87

Claremont Road 193, 236, 239

Clavering, Lady Jane (1669–1735) businesswoman 48, 60

Clayton, John (1792–1890) Town Clerk 3, 94, 103, 150

Clayton, Nathaniel (1754–1832) Town Clerk 103

Clayton Street 94, 95, 213, 220, 232

Clifford's Fort, North Shields 192

Cloth Market 11, 78, 161–3

Cloth Market area (Goad's Plan) 160

Coastal Detail of Sheet 78, Newcastle upon Tyne (OS, 1947) 194

Coatsworth, John, artist 222–3

coal measures xi, 27

coal mining 24–9, 102, 110, 114; collieries: Benwell 11; Blackburn 60; Blyth 20; Elswick 11; Fawdon 60; Heaton 63; Longbenton 59; Marsden 75; Ravensworth 60; Seghill 60, 63; Tanfield Moor 60; Templetown 75; Westhoe 75; Whitburn 199

coal trade 36–7, 39–41, 55, 66, 71, 85, 102, 133–5

coal transport 23, 26, 29, 59–63, 113

Coket (Coquet) Island 19

Colby, Thomas (1784–1852) Director of Ordnance Survey 99

Collingwood Street 79, 90–1, 162; fire 128

Collins, Greenvile (1643–1694), naval officer 20–3

Commissioners for Tidal Harbours, report 132

Common, Jack (1903–1968) socialist writer xii

commuting 107, 195, 198

Condercum, Roman fort 3

Conservancy Commission 132

conservation, historic buildings 208, 219–20

Consett Ironworks 135

Conway, Lord Edward (1594–1655) Royalist commander 36

Cookson, John & Company, glass manufacturer 75, 86

Cooper, Robert, surveyor 100

Coquet Street 229

Corbridge, James, surveyor xii, 46–9, 93; map (1723) 32, 53

Corn Exchange 11, 153

Cossins, John (1697–1743) cartographer 47

County Borough of Tynemouth Map (1948) 198

County Courts (1812) 90

Covid pandemic 211, 213, 241

Cowen, Joseph Jr (1829–1900) politician 144

Cragside, Northumberland 125, 203

Cramlington, waggonway 63, 111

Crawcrook, waggonway 60

Crawley, Frederick J. (d.1966) Chief Constable 193

Crowley, Sir Ambrose (1657–1713) iron master 43–5

Crowley, Millington & Co., iron works 45

Cruddas, George (1788–1879) businessman 173

Cueste der noordt custe van Engelandt (1585) 18

Cullercoats 60, 111, 198

Cunningham, William, engraver 31

Cushie Butterfield, song 162

Customs House (1863–4) 199

Daily Leader 121

Danvers, Verney L. (1895–1973) commercial artist 150

Darlington 89, 111

Darras Hall estate 186

Dawson, Robert Kearsley (1798–1862) Ordnance Survey draughtsman 98–103

Dean Street xii, 77, 78, 79, 128

Denton Dene, proposed park 187

depressed areas 189–92, 193–4

De rivier Tyne of NewCastle geleegen aan de oost Kust van Engeland in de Noord Zee (1730) 22–3

Development plan of Ashburton Estate, Gosforth (c.1900) 205

Development Plan Review Newcastle City Council (1963) 206

Development Plans (1945) 239; (1963) 239

diseases 142–5

Dobson, Alexander (d.1854) architect 128

Dobson, John (1787–1865) architect 16–17, 89–90, 93–7, 106, 129, 207–8, 235

docks 29, 60, 75, 133–5, 191, 203, 216

Dockwray Square, North Shields 73, 198

Dodd, Barrodall Robert (c.1780–1837) surveyor 83

Dodd, Ralph (1756–1832), canal engineer 81

Doetichum, Jan van, engraver 19

Dortwick Sand 131, 133

Dove Marine Laboratory, Newcastle University 241

Doxford Park, Sunderland 232

Drayton, Michael (1563–1631) poet xi

Dunston, coal staithes 135

Durham 32, 54, 89; coalfield 26, 29

Dutch cartography 19

Dykes, Cuthbert, town surveyor 120

Easington 20

Easten, Alderman Stephen (1866–1936) builder 155

East Quayside, development 129, 215–17, 227

East Quayside Masterplan sketch (1990) 214

Edward VII Bridge (1906) 111, 247